Practical Knowledge-Based Systems in Conceptual Design

John Miles and Carolynne Moore

Practical Knowledge-Based Systems in Conceptual Design

With 42 Figures

Springer-Verlag
London Berlin Heidelberg New York
Paris Tokyo Hong Kong
Barcelona Budapest

John C. Miles, BSc, MSc, PhD
Carolynne J. Moore, BSc, PhD

School of Engineering, Division of Civil Engineering,
University of Wales College of Cardiff, P.O. Box 917, Cardiff,
Wales CF2 1YF, UK

ISBN-13:978-1-4471-2044-5   e-ISBN-13:978-1-4471-2042-1
DOI: 10.1007/978-1-4471-2042-1

British Library Cataloguing in Publication Data
Miles, John C.
  Practical Knowledge Based Systems in
  Conceptual Design
  I. Title   II. Moore, Carolynne J.
  745.40285
ISBN-13:978-1-4471-2044-5

Library of Congress Cataloging-in-Publication Data
Miles, John, 1950–
  Practical knowledge based systems in conceptual design / John
Miles and Carolynne Moore.
    p.   cm.
  Includes bibliographical references.
  ISBN-13:978-1-4471-2044-5 : $90.00 (est.)

  1. Computer-aided design. 2. Expert systems (Computer science)
I. Moore, Carolynne, 1966–   . II. Title.
TA345.M54 1993   92-46682
620'.0042'0285—dc20   CIP

© Springer-Verlag London Limited 1994
Softcover reprint of the hardcover 1st edition 1994

Typeset by Asco Trade Typesetting Ltd, Hong Kong

69/3830-543210 Printed on acid-free paper

Contents

Preface

Artificial Intelligence (AI) is not a new discipline but only recently has it started to be applied to the solution of practical problems. One of the major areas in which this has occurred is the subsection of AI known as expert systems. From these have developed Knowledge-Based Systems (KBS) and indeed many would argue that the terms expert system and KBS are synonymous. What are known as "first generation expert systems" attempt to replicate human reasoning in narrow, well defined areas of knowledge. Such systems suffer from the fault that they are "brittle", that is, they fail to give satisfactory answers if the user strays, even to a tiny extent, outside the domain of the system.

The development of such systems has typically been for diagnostic and other selection type problems, but over the last decade researchers have been investigating the use of expert systems for design. A lot of work on design systems has appeared in the literature, but as yet there are no reports of such systems being in everyday use by designers.

At Cardiff we have a research team looking into the use of KBS for design. Like others working in the area we started with first generation expert systems, but unlike many research teams all our work has been in close collaboration with practising designers. Most of the systems which we have developed have been tested in design offices by designers. Thus we have developed an appreciation of what designers themselves want and how best to develop systems which will suit the design process.

At the moment therefore the technology is in a state of flux, the research community is working on second generation systems and when one looks in particular at design, a number of new ideas are emerging. Although it will be some years yet before KBS become commonplace in design offices, the technology has just about reached the stage where some applications will be viable as practical tools and so we feel that this book is timely.

The book is aimed both at researchers and practising designers. We have attempted to give a fresh perspective which does not slavishly follow the accepted AI paths but instead is based on our own experience. The methods that we describe have been tested on practical scale design problems and we are therefore confident that they are both appropriate and represent best practice.

Our work has shown the importance of understanding the human element in the design process and so this is given a substantial amount of consideration. The use of jargon has been avoided wherever possible and any obscure technical terms which have been used are hopefully fully explained. Some people take it as a sign of great intelligence to be able to describe things in terms that nobody else can follow. We do not subscribe to this school of thought.

Examples of systems which have been developed to solve specific design problems are included. In our experience people learn best by looking at examples, and so we hope that this chapter will provide useful hints for those wishing to work in this area. Although both authors are civil engineers and hence our examples are drawn from this area, the book has been written with other disciplines in mind. Design is an activity which is undertaken by many professions and although there are at first sight substantial differences between the activities of these groups, the available evidence suggests that the cognitive processes are very similar.

Finally we give some hints as to what the future might bring. All forecasting is fraught with difficulty and in an area where the technology is changing rapidly any attempts to predict the long-term future are bound to be speculative, but probably over the next five years our predictions will be reasonably correct. On that optimistic note we will end this preface.

Cardiff, 1993
John Miles and Lynne Moore

Acknowledgements

In any work of this nature there are so many people with whom one comes into contact that it is difficult to acknowledge everybody. The following includes a list of the more major debts which we owe, and we offer in advance our full apologies to anyone we have omitted.

The first debt is to Dr J. F. Miles, brother of one of the authors, who first provided the idea for this work. Then to our industrial collaborators, in particular Malcolm Fletcher, MBE, Alan Hayward, Alan Simpson, Sriskandan, Mike Gladstone, Philip Patterson and the staff of Sir William Halcrow and Partners, Mott MacDonald and Partners, Cass Hayward and Partners, Sir Alexander Gibb and Partners, Dr Colin Powelsland, Dr Neil Harkness, Water Research Centre, Wessex Water plc and North West Water plc.

In addition we would like to acknowledge the help of our colleagues and co-researchers and in particular Professor Roy Evans, Dr Chris Miles, Dr C. K. Soh, John Hooper, Brian Philbey, Steve Jones, Alistair Barclay, Dr John Brandon, Dr Ben Barr, Annthia Walker, Sheila Foley and Syd Swarbrick.

No research can take place without funding and for this we are grateful to the Science and Engineering Research Council and the Water Research Centre.

Last, but by no means least, we would like to recognise the contribution of someone who but for his enormous workload would have been a co-author, Professor Pham.

Chapter 1

Knowledge-Based Systems: An Overview

Aims of the Book

Our reasons for writing this book are that for some years we have been building Knowledge-Based Systems (KBS) in conjunction with practising designers. During the development of these systems it has been found that many of the "textbook" approaches for the development of KBS are less than ideal for design systems. To enable useful software to be developed we have had to devise our own methodology. Having been through this time-consuming and demanding experience, we felt that it would be beneficial to pass on the techniques which we have developed to others who wish to work in this area.

The aims of this book are therefore:

- To show how KBS technology can be applied to develop tools which are of use to designers, particularly for the conceptual design process.
- To approach the problem of KBS development from the stand-point of producing systems which primarily address the needs of practising designers rather than the research community, but in so doing put forward techniques which will enhance the relevance of research.
- To base the book on methods and techniques which we personally have either evaluated or developed. This gives the reader confidence that our techniques are both relevant and successful.

From the above it can be seen that our aim is for the book to be of use both to practising designers who wish to know about KBS and researchers who wish to work towards the development of improved techniques which will increase the power and productivity of design systems.

To date, so far as design is concerned, KBS are a relatively new and untried form of software system. A large number of research prototypes have been constructed, some of which are covered by the References section at the end of this book, but there are few, if any, design KBS in everyday use. However, the technology has recently started to reach the stage where it can be of some use, although further research is needed before the full potential of such systems can be realised.

What benefits will KBS bring to the design office? Obviously for a detailed answer one has to read the whole book. However, the potential benefits can be summarised as follows:

- KBS will allow easier and more structured access to design expertise which will result in a more informed workforce and hence better designs.
- The implementation of KBS and related software should result in some savings in design costs, although it must be emphasised that with current technology and for the foreseeable future, design KBS will not replace designers but instead support them.
- The quality assurance aspect of design KBS will produce major benefits. The memory power and consistency of computers can be used to check designs and help to ensure that all the relevant aspects have been considered.
- Experience and "know-how" are important factors in the design process. They cannot be costed and so they cannot appear in any economic justification for using KBS. However, one of the major benefits of such software lies in the retention of and easy access to such knowledge.

In the following paragraphs we initially look at the use of computers in design before going on to define what KBSs are and to look at further aspects of their relevance to the design process.

Computing Systems and Design

For several decades, engineers have made increasing use of computer software to help them understand and analyse the behaviour of the artefacts that they are responsible for designing. The software systems employed have largely been based on well founded mathematical and physical concepts.

For design, such systems are only of use for analysis, this being a relatively limited part of the entire design process. The success of this analytical software has been based on the fact that computers are able to store and process large amounts of numerical data at high speed – a task which humans find difficult. For example, software exists which permits the analysis of complex groundwater flow problems (Rushton and Redshaw, 1979). More recently the graphical aspect of design has been enhanced by the introduction of CAD systems.

However, design consists of much more than analysis and graphics. As is shown in Fig. 1.1, engineering design can be broadly idealised as consisting of three components which are largely independent. The area which to date has been resistant to the introduction of computer systems is conceptual design (sometimes called preliminary design). However, the development of Knowledge-Based Systems (KBS) has enabled software developers to start to provide systems which are relevant to the needs of those undertaking conceptual design. This is the main subject area of this book because this is where our experience has occurred. It should, however, be made clear that KBS do have uses in other areas of design and much of what is contained in this book will be of relevance to those wishing to work in these areas.

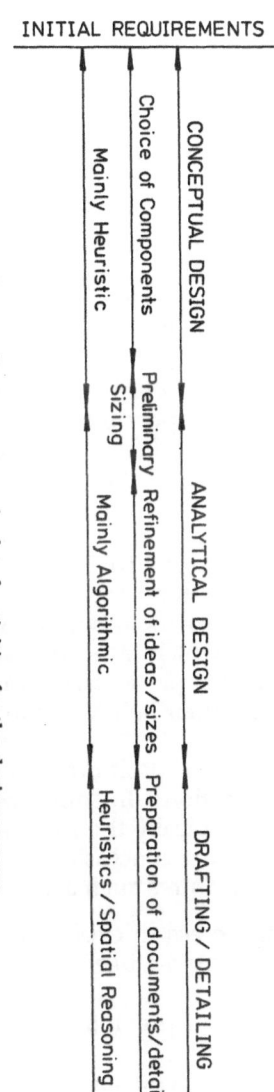

Fig. 1.1 Components and related activities for the design process.

What are KBS?

The detailed answer to the question as to what constitutes a KBS is given in the following chapters, but it is useful at this stage to provide a brief introduction. As with many topics, it is easier to state the features which a KBS does not possess. For example, with a typical numerical/ algorithmic computer system, the software is normally used to provide an answer to a well defined problem such as the stresses and strains in a component, or pressures and flows in a hydraulic system. There is gener- ally only one possible solution to the problem although the value of the solution will vary depending on the input. The solution process used by the software follows a well defined route, often with input data from a pre-prepared data file, and provides a single answer to the problem (although the answer may have many sections). Often the only interac- tion between the user and the system is to create the data file, set the program running and collect the results. Such systems rarely contain help facilities for the user.

In contrast to this, a typical KBS will be used to solve an ill-defined problem to which there are many possible answers. So, for example, rather than finding pressures in a fluid flow system, it is more likely that a KBS would be used to say what form the system should take. As there are many possible outcomes (although usually only a few feasible solu- tions), the system will have to sort through its knowledge base to find a suitable answer or answers. In a typical KBS, this will involve interaction between the user and the system, the user being asked for information about the problem. The KBS will use its in-built knowledge to ensure that it only asks for the information which is needed to solve the particular problem being dealt with. This is a particularly attractive feature of such systems; they are able to modify their behaviour to suit the requirements of a given problem.

As well as the KBS asking the user for information, the user can ask the KBS for help with answering questions. In some cases the KBS may even suggest a suitable answer. Also typically the KBS will explain the reason- ing process used to reach a particular conclusion.

Many KBS contain additional features such as:

- Glossaries which users can access to explain words with which they are unfamiliar (this is particularly useful for jargon and technical terms).
- The ability for users to change a previous answer if they realise they have made a mistake.
- A facility which enables users to type new knowledge into the system. This latter feature can, however, cause problems of clashes between existing and new knowledge.

Although it is possible to develop KBS using established computing lan- guages such as BASIC or PASCAL, in some cases the effort required and the resulting complexity of the system is such that it is far better to use specialist software. There are two well established languages for KBS: PROLOG and LISP. In addition to these there are a large number of software packages of varying levels of complexity. The choice of software for any particular problem is always a difficult decision. We tend to use the following guide-lines:

- If the system is relatively simple and unlikely to undergo further development then one of the standard shells or the more simple toolkits will suffice.
- For more complex problems or for those which are likely to evolve with time, it is far better to use a less restrictive form of development software such as PROLOG or C.

The choice between a procedural language (e.g. languages such as BASIC, PASCAL, C and FORTRAN) or an AI language (e.g. PROLOG and LISP) is usually made on the basis that, if the problem is not one in which combinatorial explosion is likely to occur because of the number of possible solutions, then one can use either form of software. If the number of possible solutions is very large then one should only use an AI language. The latter are particularly efficient at searching large knowledge bases. Also, it is far easier to cope with non-numeric knowledge using AI languages.

A common feature of KBS software is the similarity of the knowledge bases to natural language so that large parts of the code are intelligible to people other than experienced programmers. In the long term this holds out the attractive prospect of programs which can be readily modified and updated by users. The current difficulty with this is that allowing users access to the knowledge base of a system can easily lead to the corruption of the system logic. As is described below, the acquisition and verification of the information which forms a knowledge base is a process that requires a lot of hard work and care. As yet we do not have the technology to ensure that user additions or modifications to the knowledge base will not clash with or corrupt the system.

Historical Perspective

It is useful to consider briefly the development of KBS and Artificial Intelligence (AI) so that the problems faced in developing engineering design KBS and the associated jargon can be placed in their proper perspective. The development of AI is of some interest because although it has existed as a research subject for roughly the same amount of time as, for example, finite elements, AI techniques are only just beginning to make an impact in design whereas finite elements are in everyday use.

As with many aspects of human endeavour it is difficult to identify exactly when AI first came into being as a discipline, but it is generally agreed that the main development has occurred since the 1950s with much of the work taking place at Carnegie–Mellon University under the guidance of Newell and Simon. Various development phases can be traced. A good summary of this development can be found in Forsyth (1989a). For a considerable length of time, the thrust of the development was largely directed towards simulating human intelligence. During this phase there were some successes in very limited areas, for example, the checker (draughts) playing program of Samuel (1963) and similar chess programs, but AI remained largely a research interest of such people as cognitive psychologists and computer scientists.

Gradually, however, several practical applications of AI have emerged such as robotics, machine vision and expert systems. The latter formed the basis of what are now known as KBS, and they were developed from the idea that what human experts seem to possess is "know-how". This consists of such things as rules of thumb which have been based on the observation or experience of a large number of case histories. Thus expert systems were developed as simulations of human intelligence but in very narrow and limited domains (i.e. areas of knowledge).

The first true expert systems was DENDRAL (Feigenbaum et al., 1971) with the next major advance being the development of the medical diagnosis system MYCIN (Shortliffe, 1976). MYCIN contains many of the features which have become an accepted part of expert system technology, such as the ability to explain the reasoning processes which have lead the system to reach a particular conclusion.

From such beginnings, aspects of AI have gradually begun to be applied to the solution of real-life problems as opposed to being pure research tools. This gradual acceptance has brought with it the attention of a far wider public. This awareness has highlighted some problems with the jargon of AI. For example, it has been found that the name "expert system" causes enhanced expectations. The uninitiated believe that such systems should be truly expert. While within the limited domains to which their knowledge bases apply such systems may be partially able to simulate expert problem solving, unfortunately this is not true once the system is asked to solve a problem which differs slightly from its programmed knowledge. However, in such a situation one would expect a real expert to cope. Human experts possess abilities such as common sense and the capacity for original thought, which are currently impossible to replicate adequately in KBS. In many ways the more that one investigates the use of computer systems to simulate human decision making, the more one comes to realise just how tremendously capable humans are.

The enhanced expectations which the name "expert system" gives has led to the name "knowledge based systems" (KBS) being used in preference. Inevitably KBS technology has advanced to the point where such systems can be more than just simulations of limited aspects of human behaviour and so are more than just "expert" systems. KBS developers have learnt that their techniques enable them to create systems which can complement human skills so that the combination of KBS and user produces something which is far more powerful than the sum of the parts. This is directly analogous to what happened when analysis software was introduced into the design office. The capabilities of these programs has allowed the designer to analyse very complex situations and so develop new knowledge which could not have been obtained without the software. Similarly the development of KBS will result in better conceptual designs, leading to improved products.

We do not pretend that KBS will be the only type of system that will have an impact on conceptual design. Designers make extensive use of past designs as a starting point for new schemes. If large amounts of information on past designs need to be processed then database techniques could be more appropriate in certain areas. Soh (1990) has already shown that by coupling database and knowledge base techniques one can produce very powerful design systems. There are strong similarities between database and knowledge base programming, and it is likely that in the future the boundaries between these two types of software will become increasingly blurred. However, for the purposes of this book we will concentrate on KBS.

Types of KBS

It is possible to place the KBS developed to date into one of several broad categories. For example, the MYCIN system which was quoted above was developed for the diagnosis of blood infections. Many of the KBS which have been subsequently developed also solve diagnostic problems. A good example of an engineering diagnostic system is BREDAMP (Allwood *et al.*, 1988) which can be used to help to identify the likely causes of dampness in buildings.

A KBS can be used as a "front end" to other software systems. For example, using the software developed as part of the MYCIN project, Bennett and Englemore (1979) produced a "front end" called SACON for a large structural analysis package called MARC. The latter is a very complex finite element analysis package used to simulate the behaviour of structures under various loading conditions. The complexity of MARC meant that it typically took a year of experience to learn how to use the system fully. The purpose of SACON was to act as an aid, helping the user to set up MARC correctly. As reported, SACON and MARC were separate entities but subsequently such front ends have been coupled to analysis packages. A sophisticated version of this type of KBS is the work of Soh (1990), reported in Chapter 9.

Design systems are recognised as being a separate type of KBS, possessing attributes that rarely occur in other types of system. In diagnosis, for instance, for a given problem there is generally a single correct solution (although this may involve more than one factor, for instance the dampness in a building may be due to a leaking water pipe and a failed damp-proof membrane) and difficulty occurs with the identification of that solution. In design there are many possible answers, and the challenge is therefore to determine the most suitable solution(s). This is an important philosophical distinction and it is generally accepted that because of its open-ended nature, design is one of the more difficult areas to which KBS can be applied.

It is useful to identify the capabilities of design KBS, but first it is necessary to say something more about design itself.

Brown and Chandrasekaran (1985) identify three broad classes of engineering design, these being:

- Class 1: Inventions or new products.
- Class 2: Products involving substantial innovation.
- Class 3: Design which proceeds by selecting from previously known sets of design alternatives. However, the problem is still too complex for it to be achieved by referring to existing products.

While class 3 designs may be complex, at no stage is there an open-ended problem and so the solution path will always be known. Innovation is required for class 1 and 2 designs. There are claims that in very limited areas computers have shown themselves capable of original thought (Lenat, 1982), but these claims are tenuous and it is safe to assume that, at present and for the foreseeable future, original thought is beyond the capabilities of the computer.

Consequently, given current technology, class 1 and 2 designs are beyond the capabilities of KBS but class 3 is something which can and is beginning to benefit from the development of design systems.

Chandrasekaran (1983), in an attempt to produce a taxonomy of problem- solving types, identifies a fourth type of KBS which he classifies as consequence-finding systems. Such systems reason about the consequences of contemplated actions. So, for example, if one is the manager of a complex manufacturing plant or machine, such a system would enable "what if" situations to be explored. For example, "What will happen if valve A is closed when the boiler is under high pressure?" (Chandrasekaran, 1983). The financial sector is beginning to use such systems to assess the consequences of such actions as loaning money or applying a new business strategy.

Thus four types of KBS have been identified above, these being:

1. Diagnostic Systems
2. Front End Sysyems
3. Design Systems
4. Consequence-finding Systems

This classification does not preclude any one system from having more than one capability. For example, the work of Soh (1990), although it can be classified as a front end, includes an extensive design KBS and a limited ability to investigate the consequences of changes to the suggested solution.

In addition to the above, it must be recognised that any system is unlikely to be a truly independent entity. This will increasingly be the case as KBS technology matures. Already there is a well established body of thought regarding co-operating KBS (see, for example, Huang, 1990), and likewise links between KBS and other types of system such as databases are the subject of a great deal of work.

If the maximum benefits of computers are to be realised then it is essential that systems be designed to allow links with other software to be easily set up. Without this, the transfer of information between separate systems is cumbersome and a source of possible errors.

In subsequent chapters the discussion concentrates largely on KBS design systems, but it is important to remember that the power of such systems can be greatly enhanced by links to other software.

Successful KBS Applications

One of the first commercially useful KBS systems was XCON, which was developed for determining the configuration of computing facilities by the Digital Equipment Corporation (DEC) (McDermott, 1982). The problem is complex and difficult to solve, and DEC recognises that XCON does not always give a correct solution. However, with a computer system, once the error is detected and corrected, one can be sure that the system will never make that mistake again. Such a level of performance could not be guaranteed with a human expert.

Another early success which is commonly reported in the literature is the PROSPECTOR system, which it is claimed (Gaschnig, 1980) has identified previously undiscovered mineral bodies, although some of the claims are disputed (Partridge, 1988).

The use of KBS to solve practical problems is increasing rapidly. Chung and Inder (1990) estimated that, in 1989, the Japanese had 144 KBS in everyday use with this figure representing an increase of 750% over the number in use in 1987. Figures for other countries are not available but the rate of increase is probably similar. Commercial applications of KBS are often not reported because companies do not wish to lose their competitive edge by revealing their success to rival concerns. It is known that the Shell Oil Company has several major KBS applications, one of which, despite costing several hundred thousand dollars to construct, had a payback period of only a few months.

It would appear that most of the systems in use are small, taking from a few weeks to a few months to create and test. Such systems may only contain a relatively small amount of computer code and solve a relatively simple problem but they fulfil a useful role. "Expert systems" do not need huge knowledge bases, and they do not even have to perform tasks which one would normally associate with an expert.

As well as the successes, there have been a large number of systems which have failed to develop into practical software systems. For instance, the MYCIN system (Shortliffe, 1976) contained a comprehensive range of facilities including a "natural language" interface, but the system never developed beyond being a research tool. The software was later divorced from the knowledge base to form a so-called shell (this is an expert system with no knowledge in it, the idea being that the developers of a new system can insert the knowledge from their particular domain) but this too failed to yield practical applications.

As discussed above, the AI community has a background which is not based in practical problem solving. It would seem that this has influenced the approach taken by many to the development of KBS. Too much emphasis has been placed on adding the features that the AI ethos dictates are necessary rather than creating systems that fulfil a useful, practical role. It is interesting that AI practitioners have a "wish list" of features that all KBS should contain. The ideal KBS is expected to do the following (Davies, 1982):

- Solve the problem.
- Explain its reasoning.

- Learn from completed cases.
- Restructure its knowledge.
- Break rules (when and wherever necessary).
- Determine the relevance of acquired knowledge.
- Degrade gracefully upon reaching the limits of the domain.

The list is compiled by looking at how a human expert performs and then translating these activities into suitable long-term goals for KBS. All of the above capabilities are very desirable but many of them are not achievable with current technology: indeed most systems do not get beyond the second item on the list.

Nevertheless, lists such as this frequently appear in KBS publications. Ostensibly, such aims are very laudable and all disciplines should have long-term goals, but at times one has the impression that too much emphasis has been placed on the features in the KBS rather than on the utility of the system.

Before leaving this topic it must be stated that much has also been gained from the AI approach. For example, some KBS have excellent user interfaces with well designed help facilities. Hence there is no need to spend a week ploughing through an unintelligible manual before one can use the software. Also the ability of KBS to explain to the user how a particular conclusion has been reached greatly enhances the utility of such systems, helping the user to learn the reasoning process of the domain. Such software is helping to raise user expectations and hence the standards of current software.

Will KBS Freeze Expertise? – The Deskilling Argument

One commonly voiced fear of KBS is that their introduction will lead to humans depending entirely on KBS to perform a given task. As KBS are incapable of original thought, the worry is that such systems will prevent future development of expertise in that area. It is feared that mankind will effectively be permanently locked in the state of knowledge that existed at the time the particular KBS was developed.

If such a situation is extrapolated, one can imagine historians in a couple of centuries time analysing our progress in, for example, the development of the microchip. Their report would be something like "two centuries ago mankind was progressively developing faster and faster electronics until on the 23rd August 1995 Mr/Mrs X released his/her KBS. Since that date all development work has ceased".

When one looks at it in this light the fear of freezing knowledge is patently absurd. Humans are by nature inquisitive and they learn things very quickly. Any well written KBS should contain facilities which explain the reasoning processes used, and therefore the curious user can use these features initially to gain the same expertise as the system. From then on, further experience will only help the human expert to gain greater expertise than the system. Indeed it is probable that by making the more obscure and elusive types of experiential design knowledge readily available to inexperienced designers, their progress will be enhanced. Surely the provision of high-quality information must be of benefit? At the moment, because the mechanisms for passing expertise from one designer to another are very poor, inexperienced designers have to spend a lot of time finding out how to undertake relatively routine tasks. In effect they are "re-inventing the wheel".

In addition to the above worry that KBS will stunt innovation, others argue that they will merely retard it. This is a more difficult case to counter. Possibly the best answer is that at present there is no evidence to either prove or disprove this argument.

It has to be recognised that there will be some designers who will not use the explanation features of the KBS but will just blindly accept its answers. However, with current practice we have the situation where experts are rare. Why? Because only certain people will bother to ask questions, to think about things or in the future to use the explanation facilities. There will always be people who have the intelligence and the drive to progress faster and further than their peers. KBS will help both the high-level performers and those who are less well motivated.

If one looks at analogous situations, when an expert writes a book (a much more cumbersome source of information and expertise than a KBS) this does not freeze expertise. However, there are those who will slavishly follow the book and those who will understand its contents, think about them and progress to higher levels of expertise. There is little in life that is totally new.

For those with initiative and intelligence, KBS will form a platform from which they can progress. For others KBS will form a ceiling, an upper limit on their abilities, but it is probable that such people would only make limited progress regardless of their environment.

Attitudes to "Computer Mistakes"

Although it is slightly out of place here, it is instructive to look briefly at people's attitude to computer systems and the results of using such software. When failures occur in products and those failures can be traced back to incorrect use of a computer system or a fault in the software, the tendency is to criticise the use of computers in general and say that they have led to a decrease in design standards. This is despite the fact that everything about the computer and its software was designed, written and built by humans, and so any mistake made by a computer is ultimately a mistake by a human. As a contrast to the above, if a human designer makes a mistake, a completely different reasoning process is applied to that which occurs when a computer error is detected.

The major problem with computer systems and KBS in particular, and this is in contrast to humans, is that computers do not possess the vast reservoir of knowledge that is referred to as common sense. So if a computer system produces an answer that is patently wrong, unless programmed to detect the error, the computer has no way of judging the validity of the answer. The implication of this for design KBS is that such systems should never be used to provide a final answer. There must always be a check by an experienced designer to look at the solution proposed by the system (although the designer is not infallible). The need for checking and human involvement indicates that probably the best form of design KBS is one which complements the skills of the designer, taking over in areas which humans find time-consuming or boring and leaving those activities at which man excels to the designer.

Will KBS Make Experts or Others Redundant?

Those who are unfamiliar with KBS tend to fear that the development of this type of system will lead to redundancies, either by personnel being replaced by KBS or by the contents of the knowledge base of a KBS becoming available to competitors. Indeed we have experienced a situation where a company refused to collaborate in the development of a design KBS because it was concerned that the system would make all its expertise available to its competitors and clients. This level of belief in the current capabilities of KBS is both flattering (in their estimation of the skills of the system developers) and erroneous.

Possibly expectations about the powers of KBS are unconsciously raised by familiarity with science fiction where the super-powerful, highly intelligent computer is an essential part of any plot. Such expectations are further enhanced by the use of such names as "expert system" which to the layman suggests a level of performance equivalent to an expert.

In reality there are many things that humans do supremely well which are very difficult to achieve on a computer. For example, when a Civil Engineer is asked to design a bridge, a visit to the site of the proposed bridge yields a lot of information, much of which is acquired effortlessly. If the journey to the site is down narrow constricted roads, then from the outset the engineer will know that site access may be a problem. Once at the site, a quick look around gives an excellent appreciation of topography and spatial relationships. The latter can always be supplemented by using maps or remotely sensed data. To get a computer system to take in, comprehend and make use of such a mass of information would be a major achievement.

So for developers of KBS, it is essential to recognise that to try and supplant humans in many areas of activity would be a waste of time. The resulting systems would, given current technology, perform at a lower level than humans and hence would be unacceptable.

There are, however, areas where computers can easily out-perform humans. One obvious example is undertaking a large number of repetitive calculations. We have already discussed the use of such things as finite element analysis in design. Such software is capable of analysis in much greater detail and for far more complex situations than can be achieved by humans. Interestingly, the use of such software has not resulted in any obvious redundancies among designers but rather it has increased their powers of understanding. This is a good historical example of the introduction of sophisticated design software, which has not resulted in re-dundancies despite the fact that engineers formerly spent weeks under-taking calculations.

It is impossible to say that the use of KBS in design offices will never lead to engineers being supplanted by such systems, but given current ex-perience it seems unlikely. Instead KBS should result in a better trained and educated workforce who have easy access to information and experi-ence regarding design practices and constraints. Also if KBS are de-veloped in areas in which human intelligence is slow or inaccurate (as has happened with analysis) then the combination of designer and KBS should result in a more powerful team producing better products.

What will be the Benefits of Design KBS?

Many of the benefits of using KBS in a design environment have already been mentioned, but it is useful to pull them all together and discuss the topic as a whole before moving on to the detail in the following chapters.

To any design team, or individual designer, experience is a valuable asset. If we focus on conceptual design, the processes used are largely of a nature that cannot be taught in a classroom or by books. There are those who would disagree with this statement but we will deal with the argu-ments in detail in Chapters 2 and 4. Conceptual design knowledge is largely acquired by practice and by sitting alongside colleagues and watching their performance. By this process one can gain enough in-formation to become competent, but true expertise is the product of many more years of experience. With well developed KBS it is possible to cap-ture some (by no means all) of the design expertise of a gifted individual or individuals and make it readily available to others without the need for an expert to be available.

Expertise is something which resides within the person who possesses that knowledge. Thus if that individual leaves a company, then the exper-tise goes as well. If the expertise has been captured in a KBS, then the loss is mitigated.

Moving on from this, as is shown in Chapters 2, 4 and 5, the expert is often unaware of the range and scope of his/her knowledge. By going through the structured knowledge elicitation and interpretation process necessary to formulate a KBS, the expert's knowledge can be made avail-able in a structured and well documented form. This can often be of be-nefit to the expert as well as the potential end user of the system.

The presence of good quality design KBS within design offices will lead to information about the design process being readily available, thus lead-ing to better designs and better products. Ideally a combination of database-type systems containing information on existing designs and KBS with knowledge about the design process, coupled to CAD and analysis packages, should provide the designer with a great deal of assist-ance at all stages. If used sensibly, design KBS should also offer a means of closing the feedback loop between the designer and manufacturer/ constructor. Although in many industries this is not a problem as the two work together, in the construction industry there are often only very weak links between the design office and the construction site and so any on-site problems caused by bad design tend not to be transmitted back to the designer.

A well formulated design KBS can, as well as providing design assistance and advice, also act as a check-list, ensuring that when considering a new artefact all aspects are taken into consideration. If the designer happens to forget some check or criterion then the system should provide a reminder.

In Chapter 2 the problems of bias in human decision making are covered. All humans envisage themselves as being rational people who base their judgements on sound, well thought out principles. In reality it comes as something of a shock to find that well established psychological research shows that this is often not the case. For example, humans are prone to "recentism". That is if the bias is removed, the problem has been corrected for good. Mechanisms for reducing the likelihood of bias in knowledge bases are given in Chapter 5.

So far the benefits described have been those which would be gained from a typical expert system, that is a replication of an expert designer's thought process. Design KBS are, however, capable of more than just mere simulations. We will show later that it is possible to construct systems which will exceed the capabilities of designers and thus give them the ability to examine features in more detail. The long-term future of design KBS lies in complementing the power of the designer, so that is placing undue emphasis on recent experiences. A computer system could be prone to bias if it is based on a snapshot of a single human's thought processes at a given time. However, the advantage of a computer system is that the combination of design KBS and designer leads to better designs. Hopefully this can be achieved without deskilling the design process.

Chapter 2

Knowledge

Objectives

This chapter aims to:

- Discuss the complex subject of knowledge, suggest ways to classify it and decide how it affects the design process.
- Look at heuristics: defining what they are and how they affect expertise.
- Discuss the classification of heuristics and suggest some of the benefits of heuristic replacement in KBS.
- Describe the subject of bias and illustrate its effect on expertise.

Introduction

To be able to build useful design KBS, it is first necessary to understand what knowledge is and how it impinges on the design process. In this and the next two chapters the importance of knowledge and its relevance to design are covered.

Within this chapter the subject of knowledge is introduced. The intention of the chapter is to show that knowledge is a far more complex subject than is normally supposed by the layman, to describe the types of knowledge relevant to the design process and to look at the phenomenon of bias in human decision making.

What is Knowledge?

Most people would be happy to make an attempt at answering the question "What is knowledge". Humans generally believe that they understand the processes that are used to solve problems, and they would also assert that they could explain these processes. However, the truth is that there are many everyday actions and skills performed by people that are either very difficult or impossible to put into words.

For example, most adults can drive a car. When turning the steering wheel to negotiate a bend, somehow we know exactly by how much to turn the wheel. The process is rarely one of trial and error; one does not often see cars zigzagging around bends. Also the skill does not seem to be limited to the particular type of car one normally drives; it can be easily transferred to other vehicles on which the gearing of the steering system is different. How do we do it? The answer is with apparent ease, and yet to express in words the processes which are necessary to steer a car is very difficult. "I just turn the wheel" is far from adequate. By how much, at what point in relation to the start of the bend? How does one judge the curvature of the bend? What if as one approaches it is not possible to see all of the bend, how does one choose an appropriate road speed?

For such a simple everyday operation we go through a quite complex set of judgements, and yet we are unable to express such a common skill in words. This is because of the way that the process is learnt. We learn to drive not by reading books and acquiring factual knowledge but by experience behind the wheel of a car. This is recognised by the fact that competence to drive is judged largely on ability behind the wheel rather than by written examination.

From this simple example it can be seen that certain types of knowledge are gained by experience rather than by formal learning through lectures and books. Knowledge obtained by experience is called "procedural knowledge".

Designers make extensive use of procedural knowledge. It is this which differentiates the experienced designer from the raw graduate. If effective design KBS are to be built, the system developers need to be able to access this procedural knowledge and yet, as explained above, it is not open to verbalisation. The available methods for acquiring procedural and other forms of knowledge are covered in Chapter 5.

The discussion so far may seem as if it is straying into the realms of the psychologist and many would say that this is not really something that concerns designers, particularly engineers. From a traditional point of view this is correct, because so far the software tools used by engineers have been largely algorithmic in nature. However, to create such tools engineers have happily studied mathematics. So why is there a reaction against psychology? Probably because in their training, engineers are taught that formulae and well defined knowledge are "correct" and anything else is vague speculation: the sort of thing that a humanities graduate would get involved in (what greater insult is imaginable!).

In this type of thought process there is a major contradiction. Engineers basically are practical people who create artefacts for use by mankind. In order to do so, where possible they use the laws of physics and mathematics, but there are numerous areas where these laws are inadequate. Problems created by such a lack of knowledge are usually overcome by a variety of empirical techniques. Furthermore, starting with a blank piece of paper to create a new artefact often requires the use of a considerable range of skills which lie outside the realms of mathematics and physics, and yet engineers seem to use such knowledge subconsciously. This contradictory thinking also permeates deeper into society. Engineers are not thought of as being creative people despite the fact that we are surrounded by things which have been designed by engineers.

When reading this book, the recognition that engineering contains much more than just equations and that design is much more than analysis is essential. The whole basis of KBS is that they provide a method of solving problems using computers without the need to express things in terms of equations.

In the following sections of this chapter, first we discuss the various types of knowledge and describe how such knowledge is acquired. Next heuristics and the concepts known as deep and shallow knowledge are discussed, as are the ways in which humans use knowledge. Finally bias in decision making is covered.

Uncertainty, which many regard as a vital part of KBS technology, is covered in Chapter 3.

Types of Knowledge

A full understanding of how the human brain works, how it acquires, filters, accepts, stores and retrieves knowledge and how it uses this knowledge to make judgements is a complex subject, and one which is far beyond the scope of this book. Those who wish to know more about this should study works such as those of Newell and Simon (1972), Lindsay and Norman (1977), Anderson (1985) and Sandford (1985). However, a simplified representation of knowledge as it applies to design is essential for understanding the succeeding chapters.

It is generally agreed that knowledge can be split into two major categories these being:

- Declarative knowledge.
- Procedural knowledge.

Factual knowledge is said to be declarative knowledge. For example, knowing that the centre span bending moment in a simply supported beam with a distributed load is given by the formula $wl^2/8$, where w is the load and l is the length, is declarative knowledge. Likewise concepts and relationships are declarative knowledge.

Procedural knowledge is knowledge which is related to how one performs tasks and is generally acquired by experience. For instance, referring to the above example, the information which is used to steer a car around a bend is procedural knowledge.

Usually people find that they can express declarative knowledge easily, but that procedural knowledge, because of the manner in which it is gained, cannot easily be verbalised. It is shown below that real expertise mostly consists of procedural knowledge. If systems which perform at a high level are to be built, then it is necessary somehow to identify the relevant procedural knowledge and then express it in a computer code. As it is something which is not accessible to verbalisation, this process is difficult.

When learning how to undertake tasks, we generally start by acquiring declarative knowledge. So, for example, when learning to toss (i.e. turn over) a pancake in a pan, one is told that the pancake must be sliding freely in the pan and then one flips the front of the pan up, watches the pancake carefully and catches it when it has rotated through 180°. Having gained the fundamental knowledge, one can begin to practise.

The next process of skill acquisition is called the "associative stage". In terms of tossing pancakes it can involve a few mistakes, but quite rapidly one achieves a basic degree of competence. At this stage, though, careful thought and concentration are still needed because the processes are not automatic. However, with further practice one reaches what is called the "autonomous stage" in which the task is performed without apparent thought and at high speed. Watch a chef in a restaurant to see how easy tossing pancakes can be!

Application to Design

If one looks at the relevance to design of these three stages of gaining competence, the initial declarative knowledge is gained via lectures, tutorials and design practicals in college or university. Here the basics of design are taught. For engineers in the English-speaking world this tends to concentrate on how to analyse the performance of designs and the acquisition of drawing skills. The teaching of conceptual design is very limited, usually because the skills needed to undertake this stage of the design process can only be acquired by a lot of practical experience and hence are not possessed by most academics who usually have limited experience of work in industry. This is implicitly recognised in the qualifications of engineers who are required first to obtain a degree (a recognition of their declarative knowledge) and later, after some years experience as engineers, they can qualify for further professional qualifications (acknowledgement of their procedural knowledge).

The available evidence in the literature suggests that in the German-speaking world things are different. Academics are allowed to have far greater contact with industry and this seems to spill over into the teaching of conceptual design to engineers.

The teaching of architects at undergraduate level contains much more conceptual design but, in Britain at least, the structure and length of their courses allow for this.

In the early stages of learning about design, progress in undertaking any given design task is slow. There is a need to check constantly and revise basic procedures, nothing is automatic and there is no background of previous work to call on. This is why in a design office the fresh graduate takes some months before he or she starts to become an economic asset to the company.

With some practice, our novice designer begins to pick up the basic rules and learn the more fundamental concepts of the process. This is the associative stage. Nothing as yet comes automatically, but with a lot of thought and care some useful work can be turned out. Progress at this stage is rapid and after a few years the designer is of real value to the company (and relatively cheap to employ!). Then begins the long haul of gaining experience of many different situations to reach the autonomous stage. Those who perform at the highest level become experts, those who fail to progress become useful designers albeit at a lower level.

The design skills which experienced designers possess enable them to carry out quickly many tasks which to the uninitiated seem extremely complex and would take a long time to complete. This level of expert performance is achieved without apparent effort and even without a great deal of obvious thought. This is because the experts' knowledge of the design process is such that they intuitively know what to do. They no longer have to go through the mass of assessment and thought that the less experienced need to reach the same goal. Their experience enables them to achieve a very high level of performance at great speed.

During the development of a KBS for the conceptual design of bridges, Moore and Miles (1991b) worked with several expert bridge designers. Such people were able to undertake the conceptual design of a small to medium size bridge within a few hours. This would include such judgements as the aesthetics of the bridge including the harmony of its appearance with its surroundings, economics, site access problems, structural behaviour, construction scheduling, ease of construction, maintenance, traffic constraints and the client's requirements. To cope with such a wide range of factors at speed is an impressive feat and requires a lot of background knowledge and experience.

Thus design knowledge consists of both declarative and procedural knowledge. It is thought that those who perform at the highest level in terms of competence and speed tend to make extensive use of procedural knowledge, although the evidence to support this assertion is slim. Unfortunately, studies of designers and how they work are rare but it is possible to draw parallels with expertise in other domains (Newell and Simon, 1972). Some of this design knowledge is based on a compiled understanding of underlying declarative knowledge but much of it is gained from experience. This is especially true of conceptual design for which little declarative knowledge exists.

Further Classifications of Knowledge

Algorithmic Knowledge and Heuristics

In addition to the above split into declarative and procedural, there are other ways of classifying knowledge.

For example, there are two distinct approaches which are used at various stages throughout the design process – these being algorithmic and heuristic. In simple terms one can conceive of the algorithmic approach as utilising equations which are typically based on Newtonian physics, whereas the heuristic approach uses rules of thumb based on experience. The former tends to take longer but gives a much greater assurance of a correct answer than the heuristics. The distinction between heuristics and algorithms is of importance in the development of KBS and so the topic is covered below in some detail.

Before proceeding further, a few examples of each type of approach will help to clarify their advantages and disadvantages.

Simple Example

First consider the available methods for finding the roots of a quadratic equation. Take, for example, the equation

$$x^2 + 3x - 10 = 0 \qquad (2.1)$$

To obtain the roots of this equation using the algorithmic approach, one applies the standard equation

$$x = \frac{-b \pm \sqrt{b^2 - 4ac}}{2a}$$

where a is the coefficient of the squared term (in this case equal to one), *b* is the coefficient of the middle term and *c* is the last term. Applying this equation to the above – one obtains the solution that the roots are -5 and $+2$.

Using a heuristic approach, the equation can be solved by inspection. Each quadratic consists of two terms of the form $(x + u)$ and $(x + v)$ where u and v are constants. The two terms when multiplied together give the quadratic. The last term of the quadratic is therefore the product of u and v and the constant of the middle term is the sum of u and v. Using this knowledge, one can therefore solve the quadratic by inspection by breaking it down as follows:

$$x^2 + 3x - 10 = (x + 5)(x - 2) = 0 \qquad (2.2)$$

Therefore the roots are $x = -5$ and $x = +2$. Although when described in detail the heuristic method seems to be as complex as the algorithmic, with a little practice it is very simple and quick to use provided the roots are integer values. If the roots are not integers then the method requires a trial and error approach to obtain a solution.

Although it is a simple example, the above contains all the elements of the algorithmic and heuristic approaches. The algorithmic approach is more complex and requires more effort, but is guaranteed to produce an answer. The heuristic approach requires a degree of practice to obtain expertise, but once this level of competence has been reached it is very quick and easy to use provided the roots are integers. If this is not the case then obtaining an answer can in some cases be difficult. Not all heuristics are of this form. As will be shown below, they can vary greatly, with the only common feature being that they represent an approximate method, a short-cut.

Some further examples of heuristics are given below.

Examples from Design

Generally it can be assumed that conceptual design largely consists of heuristics and analytical design of algorithms, but there are many exceptions to this statement particularly in analytical design. Designers often encounter situations where the existing algorithmic methods are not capable of solving a problem and then recourse to the heuristic approach is necessary.

Take, for example, the following case from fluid mechanics. Determining the flow which a channel or pipe will carry, given a particular energy gradient, is a common problem in hydraulics. In the 19th century, in the absence of any fully validated and theoretically correct equation, Robert Manning produced an equation which was based on a statistical correlation between observed flow velocities and energy gradients. The equation was later modified by others but still retained Manning's name, and is still based on the correlation approach so that the modern equation

$$v = \frac{1}{n} R^{2/3} i^{1/2} \qquad (2.3)$$

is not dimensionally correct because the coefficient n is not dimensionless. In effect, the equation is a heuristic which relates velocity and energy gradient for a conduit of given dimensions. This latter statement may initially cause some disquiet because Manning's equation is a well known formula and the mind tends to make an automatic association between algorithms and formulae. However, Manning's equation cannot be justified using a theoretical approach and it is only valid for the range of values used for the initial correlation to obtain the equation.

In this case the algorithmic solution would be something such as the Colebrook–White equation.

So what at first seems to be an algorithm may in fact turn out to be a heuristic, although the converse of this statement is seldom true.

Next an example of a situation where the theory is inadequate, making it necessary to use heuristics. The example comes from the treatment of polluted water. One method of removing suspended solids is by allowing them to settle out under the force of gravity. This takes place in so-called sedimentation tanks.

When considering the settlement of solids in sedimentation tanks, designers always make use of heuristics based on retention times (i.e. how long on average the water will be in the tank) rather than considering the sedimentation process in detail. In cases where the solids to be settled are fairly predictable in their properties (e.g. sewage) then standard retention times can be used, these having been obtained from a wealth of past experience. For example, in temperate climates a retention time of 6 hours at average flow is often used. There is no theoretical justification for choosing such a retention time, the value having been arrived at purely from experience.

When the sediment to be settled is from a river, then a different set of heuristics is used. Each river sediment has its own properties in terms of particle size and density, and the retention times in such cases are obtained from laboratory tests. The laboratory tests give a settling velocity which is then usually multiplied by a factor of 3. This is to allow for the fact that, in a real sedimentation tank factors such as horizontal flow velocities will affect the efficiency of the sedimentation. Again, as for sewage sedimentation, there is no theoretical basis for choosing the number 3, the choice being made purely on the grounds of experience.

The algorithmic knowledge on sedimentation is based on Stokes' Law which describes the behaviour of a single spherical particle settling in a still liquid. These conditions never occur in sedimentation tanks where the water is in slow but constant movement, there are many particles and they rarely resemble spheres. Hence in its basic form Stokes' Law is of little practical use although correction factors are to be found in the standard texts (e.g. Fair *et al.*, 1971; Smethurst, 1979).

Heuristics are commonly developed by designers for their own personal use. Often such heuristics are unique to one person, indeed the nature of procedural knowledge is such that designers regularly use heuristics without being overtly aware of their existence. How one can elicit some of this hidden knowledge is dealt with in Chapter 5, but it is useful to end this section with a designer-specific heuristic.

This final example deals with bridge design and specifically beam-type road bridges (as opposed to tension structures, e.g. suspension bridges, and compression structures such as arches). During the conceptual design of such bridges it is frequently necessary to have a fairly good assessment of the final depth of the bridge deck. This allows the designer to check that the deck thickness will be sufficient to provide the required strength and also that there will be sufficient clearance beneath the soffit to meet standard vehicle heights. For this type of situation, Hayward (1990) stated that a good method of estimating the depth of the deck is to divide the span length by 20.

The algorithmic alternative to this heuristic is a complex and lengthy calculation involving the loadings on the bridge, the forces and moments induced by these loadings and the structural properties of the intended beam. Several stages of calculations may be necessary to iterate to an acceptable solution (one has first to choose a beam size and then to check that it is acceptable) although an experienced designer would probably manage to get the correct answer at the first attempt.

Again the heuristic is a rule of thumb which saves time, is easy to use but cannot always be guaranteed to give the correct solution (in this case it gives an approximate answer), whereas the algorithmic approach is time-consuming but provides an accurate solution. Note, though, that the statement that the algorithmic method is time-consuming is based on the assumption that hand calculations will be used. As will be shown below, often the use of a computer-assisted design process can lead to accurate algorithmic processes being made available to designers in a way that imposes little or no time penalty.

The Basis of Heuristics

Where do heuristics originate from and what can be learnt from answering this question? The answer is complex in so far as heuristics come from many sources, but the use of heuristics is intrinsic to the way in which humans think about and react with their environment. This intrinsic nature is an important factor in human behaviour and hence in design. As we will show below and also in Chapter 4, the apparent ability of humans to subconsciously apply heuristics has an important influence on the design process. To understand how and why heuristics are used in this way it is necessary to look at why humans originally developed them.

At any one time a conscious human is receiving a vast amount of data from his/her eyes, ears, nose and sense of touch. The total amount of information is so large that it cannot all be dealt with, and so the brain sifts out what it considers to be important and ignores the rest.

For example, when walking along a busy street it is possible to make progress at some speed, missing all obstructions without noting the details of a single person coming in the opposite direction. However, should someone make a sudden noise or gesture, then our defence mechanisms "switch on" our awareness and particular note is taken of the potential threat. Likewise, when sitting in a room listening to a conversation, the tendency is to concentrate on the speaker, ignoring all other input to the senses unless someone else does something such as opening a door or makes a sudden noise. Then immediately attention is focused on the intrusion and for a few seconds the speaker is ignored.

In effect, heuristics are used to select which information to consider. This ability to choose what to concentrate on and to make such decisions quickly is central to the success of our defence mechanisms.

Humans are supposed to have their origins on the plains of East Africa. The success or otherwise of their defence mechanisms would in those days determine the chances of survival. Thus the ability to use heuristics rapidly to choose what to respond to was vital. These days the same basic procedures are being used to cope with complicated designs.

So heuristics are something which humans use constantly to cope with the excess of incoming information, but there is evidence to suggest that this behaviour spills over into other activities undertaken by humans. For example, when designers are faced with a mass of information, they tend to reject some of the data and just concentrate on what they consider to be important. This can result in bias in decision making (discussed below) and consequent errors in the design process (discussed in Chapter 4).

KBS Heuristics

So far, the discussion on heuristics has implied a very broad definition ranging from equations based on regression analysis to intrinsic defence mechanisms. In the standard KBS texts a more narrow definition is used with a heuristic being defined as a rule of thumb developed by an expert as part of his/her expertise. In the remainder of this chapter, and indeed this book, the discussion will largely remain within the bounds of this narrower definition.

It is not necessary to be an expert to develop heuristics for a given task, but generally it is recognised that a part of what constitutes expertise is the creation and use of heuristics which enable one to solve complex problems rapidly and accurately.

The heuristics which are developed by experts are generally the product of a lot of experience; a compilation of a wealth of knowledge which can be used, often in an intuitive manner, to solve problems. When people talk about capturing expertise using knowledge elicitation techniques (see Chapter 5), much of what they seek lies in these rules of thumb.

Expert heuristics are not generally readily available to other practitioners and yet there are examples of such heuristics which have become part of commonplace design methods. The example is given above of sewage sedimentation tanks which are designed for a retention time of six hours at average flows. The origin of this heuristic is lost in the mists of time but presumably it was the product of a single expert.

An example of another type of heuristic is also given above: that of Manning's equation which was derived by using a large data set and statistical regression techniques. There are other examples of heuristics which have been derived using similar methods. For example, the work of Mackenzie and Gero (1987) looks at methods of deriving design heuristics based on the results of Pareto optimisation. The resulting rules are used for the design of one-way floor slabs in buildings.

Classification of Heuristics

In our work on the development of design KBS (Miles and Moore, 1991) we have found it to be advantageous to try to classify heuristics into groups which are related to their usage and source. By doing this one can begin to understand and analyse individual heuristics, and this then makes it possible to look for the underlying reasons as to why the heuristic was initially developed. From this one can then move forward to see if it is possible to provide a replacement method which will be more accurate and powerful. In this section we will look at the classification process and then in the following section the idea of replacement is covered.

In analytical design the computer has made some significant inroads into parts of the design process. For example, in structural design, techniques such as moment distribution have been largely supplanted by the use of computer packages which are usually based on finite element analysis. Such packages have enabled designers to analyse and comprehend the behaviour of complex systems in a manner which was not possible with hand calculations.

From the success of analysis software, some useful ideas can be gained for the ways in which KBS design systems should develop. Almost all analysis packages make extensive use of the abilities of digital computers to store and process large amounts of data at high speed. This is in contrast to the capabilities of humans who have very limited short-term memory (e.g. try and remember a seven figure number while undertaking some other task) and, for example, quickly have to resort to paper and pencil to undertake anything other than the most trivial calculation. This ability of analysis software to provide something which enhances and complements the power of the designer is the key to its success.

Initially with KBS the emphasis was largely on the "expert system" approach with systems being developed to replicate the performance of human experts. While there is much to be gained from this, the mismatch between the abilities of humans and the capabilities of computers is such that maximum use is not being made of the available potential in the combination of designer and intelligent design system.

So it is important when developing design KBS to utilise the potential of the available computing power. However, existing design methods, especially for conceptual design, have all been developed to suit the capabilities of humans rather than computers. So, for example, where problems are reached which require a lot of short-term memory and repetitive processing to achieve a solution, the human designer will tend to resort to heuristics, which give an acceptable answer quickly. If a design KBS system were available, it may well be possible for it to provide an algorithmic and hence more accurate solution, or at least to provide a more accurate heuristic by utilising the data storage and processing powers of computers.

To help with the identification of heuristics which could possibly be replaced, we have derived the following classification system. Before describing the system, however, it must be emphasised that this does not fit in with any of the taxonomies of heuristics which have been derived by psychologists, and is not based on any theoretical understanding of heuristics. Also we do not pretend that the classification system is unique or exclusive, for one could quite easily derive other systems. The approach adopted is entirely pragmatic and is based on our experience during the development of design KBS. It will be seen that it is difficult to place some heuristics into a single group and that they could, with justification, fit into at least two groups; but this does not restrict the benefits that are obtained from the identification of heuristics which could possibly be replaced.

Short-Cut Heuristics

Typically when considering heuristics, one thinks of them as being a short-cut to a lengthy process or some way of reducing a complex process (normally an algorithmic process) down to a package of work that is feasible both in terms of time and design cost.

For example, during the conceptual design of bridges, designers use heuristics to choose an economic form of construction for a given span length, so, for example, with steel bridges up to about 25 m span it is cheapest to use rolled sections (lengths above this are not readily available). Beyond this span, plate girders are the economic choice up to spans of about 35 m when it becomes worthwhile considering lattice and box girders. If a particular design falls close to the boundaries of the above span ranges (say between 23 and 27 m), or if there is a feeling that particular conditions might invalidate the above rules, then further heuristics will be used to try to more accurately identify which is the cheapest option for that particular situation.

Such heuristics are based on a wealth of past experience, information gained from tenders for constructing previous work and data gleaned from publications and personal contacts. The heuristics are quick to use and easy to implement, and therefore help to reduce design costs, which to the designer or at least to his employers is a great attraction. However, the designers themselves will admit that changes in costs of materials, special conditions which might pertain to a particular job (such as transport difficulties) and variations in the general economic climate can all influence the costs of a given form of construction. Therefore their heu-

ristics are at best guide-lines, but without a great deal of work to evaluate the various options the designers have no other tool available. So
despite the recognition of the fact that their heuristics are not ideal, bridge
designers currently have no economic alternative.

Heuristics Based on Background Knowledge

Designers use some algorithmic procedures so often that after some time
they come to know the answer to a given problem without needing to go
through the full procedure. Therefore the designer effectively develops a
heuristic to replace the algorithm. Strictly speaking, such heuristics are
"short-cuts" and thus in some ways these heuristics are a sub-set of the
above class.

The distinction between background knowledge heuristics and short-
cut heuristics is that the latter need regular updating to keep their
relevance and are at best approximate, whereas the former are based on
well founded fundamental equations which do not change with time and,
provided the designer's memory does not fail, will produce accurate
solutions.

Heuristics Based on Ill-Defined Concepts

There are many areas of design where it is necessary to make decisions
and yet no formal body of knowledge founded on well established fundamental laws exists to support them. Designers successfully cope with
this by using heuristics.

An obvious example of such an area of knowledge is aesthetics which is
a very ill-defined area and subject to the whims of fashion. However,
there are some generally accepted heuristics. For example, a widely
accepted rule in bridge design is that, from a purely aesthetic stand-point,
two span bridges should be avoided, especially if the spans are of equal
length and the abutments are large. The reason for this is that such structures lack a focal point for the eye to concentrate on. However, such
bridges are often the most economic solution for spanning a dual car-
riageway and so a compromise is used where the abutments are placed
well back from the road. This lessens the problem by placing more emphasis on the central pier and hence making this the focal point.

Heuristics Based on Empirical Data

Often when theoretical knowledge is inadequate, recourse is made to a
series of experiments which are used to provide a sufficient amount of
empirical evidence on which to base predictions. In engineering, theory
has often lagged behind practice and so such techniques have been commonplace. Manning's equation, which has been discussed above, is an
example of such an heuristic.

An interesting historical example concerns the work of Fairburn, a

Victorian engineer and a contemporary of Robert Stephenson, one of the pioneers of civil engineering. When Stephenson was looking for ways to build long span railway bridges without recourse to the use of tension structures, which are unsuitable for railways, Fairburn undertook a series of experiments which led to the development of the box girder. Several bridges were subsequently designed by Stephenson using Fairburn's results, including a bridge over the Menai Straits in Britain and one over the St Lawrence River in Canada. Interestingly enough, neither of these bridges exists in its original form!

Today, much use is still made of empirical data and many analysis methods which appear to be algorithmic in nature have elements of empiricism in them.

Heuristics Provided by Others

Design is nearly always a collaborative activity involving a group of people with complementary skills and knowledge, and so on occasions designers find it necessary to use the heuristics of others to achieve their goal. This is especially true for conceptual design where initial schemes are being drawn up and considered before the decision is being made to focus on one particular option which will be then considered in more detail.

For example, in the design of pumping systems which might be required to move such things as water or oil, the installation usually consists of a structure (known as a pumping station), in which the pumps and motors are housed, and a pipeline. The design process is usually initiated by someone such as a civil engineer who needs some initial guidance on how much room will be needed for the pumping plant. At this stage it is not possible to ask a particular pump manufacturer, because the pump supplier will subsequently be chosen by competitive tender. So an outline scheme for a pumping station is designed using heuristics which are provided on charts supplied by various pump manufacturers. The data are sufficient to enable a design to be drawn up which allows several manufacturers' equipment to fit, and then bids are invited from these people for the pumping plant and ancillary equipment. Obviously, at a later stage, the design may be modified at the suggestion of the successful tenderer to fit the actual equipment more precisely.

The heuristics in this category will inevitably also fit into one of the above categories. However, if it proves necessary to say which category, the expert who originated the heuristic will need to be traced.

Replacement of Heuristics

The purpose of the above classification system is to enable the easy identification of heuristics which can be readily replaced in a design KBS by more thorough and useful procedures. These procedures may be other heuristics which have been specially derived to suit the computer's abilities as opposed to a human's, or they may be an algorithm, but the overall aim is to produce a system which enhances and complements the skills of the designer so that the combination of designer and design KBS leads to a better product than that which would result from the designer working unassisted. If this is not achieved, then the benefits of design KBS are limited unless their use results in significant cost reductions or quality enhancements.

Using the above classification system, the type of heuristic which is most ripe for replacement is the "short-cut". These have already been identified as methods which have been developed because the problem is too complex to be economically dealt with given the available design resources.

Looking briefly at the other classes of heuristics which we have identified, the background knowledge heuristics can obviously be replaced and, for the inexperienced designer, a system which incorporates the full procedure would be useful. However, experienced designers do not need such help and if a design KBS were seen to be deskilling their job, then their feelings towards the system could well be prejudiced to the extent where the other features of the system would be rejected. This is obviously undesirable.

The best solution to this problem would seem to be a system with varying levels of interaction with the user so that the inexperienced can access a more comprehensive level of user interface than the experienced. To date we have not tried to construct such a system, nor are we aware of anyone else who has done so, but it seems to be a sensible way forward. Such a system would need to operate in such a way that the inexperienced user is able to gain sufficient knowledge to acquire the same skills as the experienced user.

The heuristics which are most difficult to replace are those based on empirical data and those based on ill-defined concepts. In certain cases it may be possible to enhance the heuristics, for instance, in the area of aesthetics, graphical representations can be of some use. It is also possible to undertake work to establish new knowledge in vague areas, but such work tends to take a lot of time and may not fit in with the resources available to the KBS developer.

For heuristics which are provided by others, it is necessary to identify which of the other four categories the heuristics fall into. The only way to achieve this is to contact an expert in the domain to which the heuristics belong.

An Example

A good example of the replacement of short-cut heuristics with an algorithmic procedure is the work of Hooper *et al.* (1992). This deals with the problem of designing a suitable disposal methodology for sewage sludge, an unsavoury but important topic!

Sewage sludge is the resulting substance left after the sedimentation processes that occur in a typical sewage treatment works. The liquid component is treated to a level where it can be returned to a natural water system (e.g. a river or the sea) without causing environmental damage. However, the settled solids have to be disposed of by other methods. The main methods of disposal are to agricultural land, to landfill sites (i.e. waste disposal tips), to incinerators or to land reclamation schemes. If the sludge is incinerated, the residual ash has to be disposed of in landfill sites.

When considering the agricultural option, the human designer has to overcome the problem of finding a suitable set of farms, within an economic travelling distance, which have sufficient capacity to cope with the expected volume of sludge. It is important to minimise the travelling distance to cut down on costs and also to minimise the environmental damage caused by lorry movements. The maximum economic travelling distance is about 60 kilometres from the sewage treatment works. Within such a distance there can therefore potentially be a large number of farms all of different sizes. To complicate matters, the cost of taking the sludge to farms varies not just with the distance to be travelled but also with the width of the access roads and the sizes of the fields. If either of the above is small, then tankers of less than the maximum allowable size have to be used and costs rise.

Therefore, as explained above, the task is to identify a set of farms which will satisfy the capacity requirement but which also provide the lowest possible transport costs. Unless the number of farms is small (and this is rarely the case given that the maximum "economic" range for sludge disposal is about 60 km from the sewage treatment works) the designer cannot provide an optimum solution within a reasonable time span and so historically the method that has been used is to look through the data on the farms, choose a set that "looks" reasonably economic and cost out that one set of farms. Obviously unless one is very lucky, the solution chosen will not be the optimum (i.e. lowest cost) and so when comparing disposal to agriculture with other disposal options, the answer will be biased against agriculture (the other disposal routes have less complexity in terms of numbers of disposal sites and are therefore easier to analyse).

Hooper *et al.* (1992) have devised a KBS to help with the strategic planning of sludge disposal. The strategic level is, in effect, equivalent to the conceptual stage in most design processes, where options are examined and evaluated to identify the disposal method which is worthy of further investigation. Initially the work of Hooper *et al.* (1992) followed the "traditional" route for KBS development of knowledge elicitation, with domain experts and knowledge acquisition from available texts, and the resulting system was then formulated so that it reproduced the experts' techniques. However, when the heuristics used for the agricultural disposal route were analysed it was realised that, by using the capabilities of the digital computer, a design method which was far superior to that used by the designers would be possible.

This optimisation problem is potentially complex and is further complicated by the fact that the farms have land areas which occur in discrete sizes, all of which are different. So the optimisation does not take place over a smooth function but over one which contains appreciable steps. Furthermore, for each potential solution, the number of farms can vary depending on whether the solution includes a lot of very small farms or a lesser number of large farms in order to provide the necessary land area on which the sludge can be disposed of.

The technique which has been chosen to solve this problem is that of genetic algorithms (Goldberg, 1989). These are based on a representation of the sort of reproduction processes that occur in natural breeding where genes reproduce themselves. Various techniques are available to ensure that new solutions are created and considered. These techniques include the development of hybrids (called crossovers) and mutations. With each generation of solutions that is considered it is necessary to have some sort of fitness function to identify the most suitable solutions which can go forward to the next generation, although to ensure species diversity some less fit individuals are also usually included. Genetic algorithms are further discussed in Chapter 6.

Genetic algorithms are a very efficient way of arriving at a near optimum solution (they are not guaranteed to reach the optimum) while only requiring a fairly limited amount of computational effort. The work on the sludge system has shown that, for a typical problem running on a 386 PC (without maths co-processor) using a program written in C, an optimal solution to the agricultural disposal of sewage sludge within a given area can be arrived at in a few seconds for a limited problem of about 30 farms and in a few minutes if the number of farms is large.

For a designer to arrive at an equally accurate solution would take many hours of work, and the chance of an error being made in such a mass of work would be substantial. So here is an example of a design KBS incorporating heuristic replacement to provide the human designer with an accurate solution within a time span that fits in with the economic requirements of the design process. This is one limited example from the many areas of design where the pragmatic analysis and replacement of heuristics could yield appreciable benefits.

For further details of the sludge disposal system see Chapter 9.

Deep and Shallow Knowledge

Some people choose to classify heuristics as shallow knowledge and algorithms as deep knowledge. The nomenclature arises because heuristics are not based (at least overtly) on the underlying laws of behaviour and are hence "shallow", whereas algorithmic knowledge is based on the underlying fundamental laws and is therefore "deep". Current KBSs are almost entirely based on shallow knowledge. Such systems can usually only solve problems which have been explicitly programmed into their knowledge base. Thus any such system which is given even a slightly different situation to solve will fail to provide a satisfactory answer. This is a major drawback of current KBS technology.

It is argued (Grimson and Patil, 1986) that only when KBS are capable of reasoning from first principles using deep knowledge will they be able to tackle problems which differ from their pre-defined solutions. The argument is taken further and it is stated that only when KBS are based on deep knowledge will true machine learning (i.e. KBS which are capable of learning) be possible although, as discussed below, deep knowledge alone will not be sufficient to achieve this.

Obviously, for many domains to which KBS technology may be applied, the state of knowledge is such that the creation of a system based on deep knowledge is not possible and as yet no KBS has been created which uses deep knowledge and fulfils the aims laid down in Grimson and Patil (1986).

However, our heuristic replacement method is immediately achievable, practical and has been demonstrated to give useful results, so although it is less ambitious in its aims than the deep/shallow knowledge argument it has the advantage that it works.

Meta-Knowledge

In addition to algorithmic knowledge and heuristics, there is knowledge about how to control procedures and methods. This control knowledge is usually called meta-knowledge (Lindsay and Norman, 1972). To a psychologist, meta-knowledge is the knowledge of one's own capabilities or so-called self-knowledge. In AI the definition of meta-knowledge is based on the psychological concept, but in practice a subtly more narrow definition is often implied, with meta-knowledge being defined as knowledge used to control the inferencing procedure within a system.

In addition to the meta-knowledge which controls the inference procedure, most KBS also contain an explanation facility which inevitably contains knowledge about knowledge, this after all being what an explanation, in part, consists of. The explanation facility provides help with answering questions and a trace of the reasoning processes used to reach a given conclusion, and is an integral part of almost all KBS. Thus virtually all KBS contain meta-knowledge at some level although when the subject of meta-knowledge is discussed in relation to KBS, it is often only in the context of controlling the inferencing procedure.

To take an example of the latter from a design KBS, Soh (1990), in his system for the preliminary design of steel off-shore structures, utilises meta-knowledge within the knowledge base to control the inferencing of the system. The meta-knowledge restricts the number of rule bases which are consulted (for a description of rules see Chapter 6) to those which are needed to solve the given problem. In this case the meta-knowledge consists of so-called meta-rules, which are embedded in the knowledge base (see Chapter 9 for further details).

As discussed in Chapter 1, one of the long-term aims of the AI community is to devise systems which are capable of learning about their chosen domain. There have been some notable attempts to create such systems, although none have been successful. Bainbridge (1988) describes some of the more notable attempts, in particular the TEIRESIAS system which contains a feature for critically evaluating any new rule which is added to the system. The meta-knowledge is held as abstract descriptions of rule sets built from empirical observations of the rules. The system is also able to examine its own data structure. However, such systems have yet to prove themselves sufficiently robust to exist outside a research environment.

Unfortunately, as discussed in Chapter 1, current KBS, although they can in many cases possess very extensive and sophisticated knowledge bases, lack common sense. Thus, unless it is explicitly told, a KBS may design a bridge deck with no supports or an electrical machine with no power input. Current KBS do not understand the reasoning processes that are encoded in their knowledge bases. As yet the techniques to achieve this understanding do not exist and until they do, then meta-knowledge will not begin to fulfil its vast potential. In many respects it is the most exciting research area for those interested in KBS.

Partridge (1988) states that effective meta-knowledge is essential if we are to have non-trivial machine learning. Once meta-knowledge has reached the level of truly understanding the knowledge within a system, then the system will be able to learn and gain knowledge and to order that knowledge in a way that will ensure that it can sensibly be used to solve future problems.

To summarise, for current KBS, meta-knowledge is a useful tool which can be used in a limited way to control the inferencing of a system and to explain how conclusions have been reached. Also if the user is to be allowed to input knowledge, then some form of control will be needed to ensure that the new knowledge is syntactically and semantically correct and that there are no logical contradictions between the existing and the new knowledge. This control will require meta-knowledge. Furthermore, if KBS are to progress substantially beyond their present capabilities, then meta-knowledge will have a major role to play.

Bias In Human Decision Making

In the section on The Basis of Heuristics, we describe how humans use heuristics to sort through the mass of information that is received by the senses, rapidly decide what is relevant and reject the remainder of the incoming data. This ability requires great skill and intelligence, and yet inevitably there must also be some problems which are caused by this data rejection mechanism.

Psychologists have for some years been studying the ability of humans to make rational, logical judgements. The results of their work are of interest to the theme of this book for three reasons:

1. They give an indication of the best way to use knowledge based programming techniques to provide truly useful design systems.
2. They show that even the most intelligent people are not cognisant of all the processes they use to solve a problem.
3. Leading on from the last point, these gaps in self-knowledge have implications for knowledge elicitation (i.e. acquiring details of the design process from design experts) as discussed in Chapter 5.

The subject of bias in human decision making is complex and as yet there are no agreed theories as to why exactly it occurs although much of the following is based on the work of Evans (1989).

A few examples of what is meant by bias will help to define the basis of what is to follow.

Humans regularly make judgements on problems from which they cannot obtain any feedback, and yet having made that decision they will try to justify their actions. For example, when someone buys a house it is a major decision and usually a lot of time is spent looking at alternative properties before the final choice is made. After the purchase has been made most people then feel happy that they have made the right choice, and yet except in exceptional circumstances they have no way of finding out whether or not their choice was correct. On a logical basis if one applied the same sort of rules that are used in scientific experiments, the only way to do this would be to live in all the other houses at the same time! This shows that often people will come to a conclusion without really thinking about the available evidence to support their thought processes.

In our lives we take many other decisions of a similar nature regarding cars, jobs, marriage partners (or just partners) and a whole host of other topics. Generally, most people feel that their judgements are correct. How justified is this feeling?

There is one common area of human activity in which judgements are made on a regular basis and of which we are all aware: the weather forecast. Weather forecasters have the advantage that they practise their skills regularly, they get rapid feedback as to the accuracy of their judgements and they have well defined theories of meteorology to help them. So unlike most of the decisions that humans make, weather forecasters have a lot background knowledge to guide them. The quoted accuracy of weather forecasts is generally between 60% and 70%. So in the light of the above discussion, are we over-confident in our ability to make good judgements?

Other ways in which humans show bias in decision making is to place excessive emphasis on particularly vivid information. Thus jurors tend to believe a dramatic eye-witness account rather than the more dry and scientific evidence presented by a forensic expert. When buying a car, despite what the consumers' guides say, we tend to rely on the evidence of one or two friends that we know who own similar cars; so that if the man who lives next door says his Edsel mark 1 is the best car ever, that tends to have a disproportionate influence on our decision.

Psychological research also indicates that people tend to place more emphasis on positive rather than negative statements, and there are some indications that people tend to ignore evidence that conflicts with their own pre-conceived ideas (how else could one explain politics?).

This selective processing of information which results in bias is thought to occur as a result of heuristics of the same sort as those which humans use to cope with the mass of data which their senses detect. As discussed above, these heuristics are used to decide which of the incoming signals to concentrate on and the rest of the information is then ignored. It is thought (Evans, 1989) that our brains also apply this selective filtering of information when making judgements, and reject some of the available information. This mechanism can, if the rejection heuristics fail to work properly, lead to bias. This selection process occurs without conscious thought and so it is difficult to control bias.

Consequences of Bias

What is more relevant to this work than conjecturing about the cause of bias is to look at the consequences of bias so far as design is concerned, and the possible implications regarding the use of design KBS.

In design, many of the judgements that are made are based on sound logical thinking, and yet it would seem probable, given the prevalence of bias in other activities, that it must occur to some extent in design. Indeed, in our own work on developing design KBS, we have seen some cases of bias.

If a designer is provided with a decision support environment which uses knowledge-based techniques, then as the designer goes through the design process, the knowledge base can interact with the designer, examine his/her decisions and where appropriate suggest alternatives. Thus the KBS will help to bring cases of bias to the attention of the designer. Also, if the knowledge base of a design KBS is found to contain bias it is possible to correct it. The correction is then permanent. It would be very difficult, given the nature of the cognitive processes involved, to remove bias from a human designer.

Conclusions

Knowledge is a far more complex subject than it first appears. Much of the knowledge possessed by humans is not readily available to verbalisation, and unfortunately some of this "hidden" knowledge is very important in design as it is this which enables experienced designers to work quickly and effectively.

Several ways of classifying knowledge have been presented including an empirical taxonomy of heuristics. This can be used to identify those heuristics which can be replaced by methods which are better able to use the advantages of computers.

Bias in decision making is an intrinsic part of human behaviour. It is therefore inevitable that it will occur in design. The probability is that this will lead to designs which do not represent optimal solutions to the given problem.

Many of the subjects introduced in this chapter will be expanded upon in subsequent chapters.

Chapter 3

Uncertainty

Objectives

This chapter aims to:

- Review the relevance of uncertainty to the development of KBS.
- Assess the suitability of including uncertainty in design oriented KBS.
- Discuss the various ways of representing uncertainty within KBS.

Introduction

Most people recognise that we are vague about many of the decisions which we take in everyday life: which car to buy, which person to employ, which colour to paint a room. The frequency of expressions like "I'm not sure . . ." or "I think that's right" illustrate how common uncertainty is. However, we somehow manage to deal with this uncertainty and reconcile it in order to reach decisions. In many cases, the reconciliation of this uncertainty takes place so quickly and easily that we are not even aware of doing it! As uncertainty affects so much of what we do, it is not surprising that it is a subject which has aroused much interest from many fields of research, including psychologists, statisticians and AI researchers. Uncertainty in AI covers far more than what is meant by the term "uncertainty" in everyday life: it covers many aspects of incompleteness, vagueness and uncertainty itself.

Uncertainty research is surrounded by debate, disagreement and confusion. Not only can people not decide whether people actually use formal methods to deal with uncertainty, they find it hard to agree whether it really matters how people deal with it, preferring to analyse how best to emulate it. Many of the debates associated with uncertainty, although interesting, are somewhat superfluous to the needs of this book. We are not concerned with in-depth discussions about the theory of uncertainty: we are more interested in whether uncertainty representation is suited to implementation in KBS development and if so, which representations should be used. Therefore, although a brief summary of some of the conflicting viewpoints will be included, this chapter concentrates on describing the various ways of representing and measuring uncertainty and discusses the relevance of including these in KBS in general and more specifically, in design KBS.

Consequently, this chapter is not oriented to the discussion of normative and performance criteria, but towards the representational adequacy of the various approaches to representing uncertainty. This is in agreement with Shafer and Tversky (1986) who say:

". . . we do not focus here on the question of whether these theories are accurate descriptions of how people think. We ask instead whether people are capable of using the theories and whether they can use them to given ends"

And

"Nor do we address here the question of whether the theories are "normative" – it seems premature to describe the use of a given tool before we have an understanding of how well and to what ends it can be used."

Therefore, in context of the rest of this book, the aim of this chapter is to look at the issue of uncertainty from a purely practical point of view and, although detailed discussion of uncertainty and its associated issues is omitted, liberal references are provided for those who wish to investigate the subject further.

Sources of Uncertainty

There are numerous reasons why we are uncertain about the information which is present or the decisions which we take. They include:

- Lack of/incompleteness of data.
- Inconsistency of data.
- Imprecision of measurement.
- Imprecision of concept.
- Combination of information from multiple sources.
- Lack of a fundamental or supporting theory.

All of these present different problems when trying to reach a conclusion and attach a level of certainty to it (Bonnissone, 1987; Jones, 1989; Dym and Levitt, 1991).

Frost (1986) illustrates these sources of uncertainty well. His examples include:

- Situations which are truly random, e.g. the motion of gas molecules or the distribution of people's heights.
- Situations which are not random but where there is insufficient data, e.g. the likelihood that a person has appendicitis.
- Situations where the knowledge involved represents an intuitive feeling.
- Situations where the terminology involved is vague; e.g. descriptions such as large, big, tall.

- Situations where the source of knowledge is not totally reliable; e.g. someone who is honest but totally absent minded.

Uncertainty in KBS Development

From the above sources of uncertainty, it can be deduced that for KBS development, the overall uncertainty which can be attached to the solution or diagnosis presented by the system can also come from a number of different sources. For instance:

- Because the input information is uncertain.
- Because the rules used to manipulate the information are uncertain.
- Because the conclusions which are drawn are uncertain.
- From the imprecision of the knowledge representation technique adopted.
- From the imprecision which is attached to the representation of the uncertainty itself!

The main problem lies in deciding to what extent these sources of uncertainty affect the problem domain being dealt with and thus whether they should be catered for in the development of a KBS. Various theories have been developed to accommodate such uncertainty, and the most prominent of these will be discussed in the following sections.

How do People Deal with Uncertainty?

As has already been mentioned, somehow people deal with apparent uncertainty and manage to make decisions in everyday life. Generally people do not employ formal or well defined techniques to enable them to deal with this uncertainty, although there are exceptions to this, such as betting on horses or weather forecasting, when probabilities are frequently used. However, when talking in terms of everyday judgements, we apparently do not use formal techniques to compensate for the uncertainty which is associated with a decision or situation.

There appear to be two fundamental schools of thought in the research world about the way in which people actually deal with uncertain information. These are described by Fox (1987) as the optimistic and pessimistic points of view. The pessimists doubt that people have the ability to make judgements in uncertain situations. The optimists believe otherwise. This debate is briefly covered here.

The Case for Pessimism

These arguments fall into four main categories:

1. People are often ignorant of particular facts or circumstances when asked to make a decision. This ignorance may be due to lack of ex-

perience, education or purely because the relevant information is not available.
2. People may have to make judgements quickly or under pressure, for instance in medicine. People tend to cope less well in these situations, as they tend to be less rational.
3. Human decision making inevitably contains bias and is in some ways inadequate compared to "correct" mathematical or logical models. Research work has shown that people do not take into account many of the mathematical factors or probabilities which should be considered, such as variance in sample size or dates.
4. People tend to be influenced by the superficial characteristics of a problem rather than giving equal attention to all the contributing factors.

Overall, the pessimist's view is that these factors preclude a person from dealing with uncertainty effectively. Therefore they feel that we should rely on accurate mathematical models for making particular decisions, as these mathematical models provide good ways of accumulating and combining evidence in order to reach a final decision.

The Optimistic View

The optimists believe that:

1. People can be educated, trained under stress and debiased so that they can deal with uncertainty more effectively.
2. People are judges against an arbitrary theory – they cope well with novel, constantly changing situations which the mathematical models do not address.

This suggests that many of the problems which people encounter are due to lack of experience or practice, and that some of these mistakes can be eliminated by sufficient training in dealing with new situations. We tend to deal with uncertainty (as we do with all information) on a dynamic basis, constantly changing our opinions to accommodate recent and contradictory information. Mathematical models can generally not deal with this level of dynamic information and consequently would be redundant in many real-life situations.

Now, inevitably these arguments are two extremes: statistical models have already proved to be effective and in many cases they can be cost effective, particularly in diagnostic situations or in commercial decision making, where a decision cannot afford to be influenced by bias, forgetfulness, recentism or irrationality. Conversely, in other areas such as design which, to be carried out effectively, intrinsically rely on the utilisation of many additional skills such as commonsense, the applicability of these models has to be questioned.

However, as has already been pointed out, an in-depth analysis of how humans do (or indeed do not) deal with uncertainty is inappropriate here. It is sufficient to say that opinions on this subject differ. In this text, we are more concerned with whether it is beneficial to represent this uncertainty in a KBS, and if so, how.

The Applicability of Representing Uncertainty in KBS

Many argue that one cannot successfully build an accurate expert system without the inclusion of reasoning under uncertainty (Jones, 1984; Cohen, 1985; Castillo and Alverez, 1991). Others maintain that the representation of uncertainty is superfluous to the needs of a KBS and that the inclusion of uncertainty merely confuses both the reasoning adopted by the system and the solution which is provided.

We tend to favour the latter category. We have not incorporated uncertainty in any of the systems which we have developed to date. This omission is a conscious decision. We feel that incorporating uncertainty is not beneficial in the domains which we are dealing with. The reasons for this decision are discussed in more detail in the following sections. We also highlight the considerations to be taken into account when choosing whether or not to reason under uncertainty in the KBS being developed.

When trying to incorporate uncertainty in KBS, there is the added complication that they are generally expected to explain their reasoning, or at least give an indication of the processes which they have utilised. To a user, an explanation which is complicated by complex statistics is often less acceptable or effective than one which is produced from basic knowledge (Hunter, 1991).

Design Versus Diagnostic Systems

As discussed in Chapter 1, two of the main groups of KBS are diagnostic systems and design-based systems. The aims and therefore the appropriateness of uncertainty inevitably differs in these two groupings.

In a diagnostic system, the answer to the problem in hand generally already exists, and can usually be found by analysing the apparent symptoms or signals. Also, there is often a single correct solution, or at least a single solution will have to be chosen. Therefore, it can be seen that in these cases it is sensible and even beneficial to attach a level of uncertainty to the solution which has been produced and to the relevant "runners up". Medical diagnosis systems such as MYCIN are prime examples of systems where uncertainty management is of great importance, as the alternative diagnoses can then be compared and considered.

In contrast, most of the domains which we have dealt with to date have been design oriented. In our opinion, design does not appear to be an area which is suited to reasoning under uncertainty for a number of reasons.

Generally, in design no standard solution originally exists: the system begins with a set of constraints from which a solution must be produced. There are no symptoms from which a cause can be deduced and there is certainly no single solution. The system can produce any design it requires, so it does not seem sensible to attach a level of uncertainty to it.

Also, subjectivity plays a large part in design, and so it does not seem pragmatic to attach a level of uncertainty to a design solution which is already recognised as being subjective in nature.

How do Your Experts Deal with Uncertainty?

If you are intending to build what we would call a "pure" expert system, that is, a system which intends to emulate expert behaviour (see Chapter 9 for further explanation of this), then you must ask yourself how the experts which you are trying to emulate actually deal with uncertainty.

In many cases, you will find that human experts find it very difficult to think in terms of probabilities or certainty factors. There is evidence of a fundamental unease and unfamiliarity with precisely quantified probabilities. This unfamiliarity can be demonstrated by experts in a number of ways:

- The expert will seem reluctant to give probabilities when asked (Speltzer and Stael von Holstein, 1983; Kidd and Cooper, 1983).
- The values produced are unreliable. Exacting values of certainty is a difficult and laborious process and the values elicited are subjective and possibly unreliable. It has been shown that the uncertainty values assigned can differ on different occasions and even at different times of the day! (Doyle, 1983; Fox et al., 1983).
- The values given are biased. This is discussed further in the following sections.

It is often found (and indeed it has been our experience) that experts do not ostensibly seem to deal with uncertainty. Instead they seem to incorporate judgements which allow for the uncertainty which is involved.

This viewpoint is also illustrated in Bramer (1990), where the systems discussed do not on the whole choose to reason under uncertainty. Many of the system developers openly admit that they feel uncertainty was inappropriate for their projects, not least because the experts which they had dealt with did not seem to be comfortable with using it.

However, some representations of uncertainty can help to overcome this problem. Alvey and Greaves (1990) have used qualifications such as "definitely", "possibly", "compatible with" etc. to quantify the level of certainty attached to a statement. They are in fact disdainful of numerical approaches, and stating that they are inappropriate, preferring to use judgements which are nearer to the expert's assessment of the situation.

Overall, we feel that if uncertainty is to be incorporated in a KBS, this informal technique may be a preferable representation approach: look at how your experts deal with the uncertainty and what representations they would be comfortable with, and chose a representation to suit. The chosen representation is then more likely to accurately reflect the uncertainty involved in the domain.

Is Uncertainty Relevant in Your Domain?

Uncertainty is not applicable in all domains, as has already been stated by numerous KBS developers. We also believe that in many cases the representation of uncertainty creates more problems than it solves.

In addition, in some domains (as in design) the uncertainty can be frequently "engineered" out of the situation by the experts so that it is no longer a problem. After all, in most cases, there is no such thing as a 100% certain design, so surely it is pointless to attach a level of certainty to the design which is produced.

A good way of judging whether or not uncertainty is relevant is to look at your experts and the domain which is being dealt with. Is uncertainty apparently used? If not, then is the uncertainty dealt with in some other way or is it just compensated for in the other expert judgements which are being made? Also, what kind of solutions/suggestions are being put forward by the KBS. As has already been mentioned, diagnostic systems are more suited to the inclusion of uncertainty than design oriented systems. Likewise, situations where a single solution must be chosen, but where alternatives are also feasible, can be enhanced by the incorporation of some sort of uncertainty measurement.

Does Uncertainty Representation Breed Uncertainty?

There is no good objective model of the way in which a human reaches a conclusion in the presence of uncertainty (Jones, 1989). Therefore, any model which is used to emulate uncertainty is at best a facet of some imagined objective model. Thus, any representation of uncertainty is only going to be an approximation of the uncertainty itself. After all, the uncertainty which we attach to a situation is generally not clearly understood or defined. Therefore, how can we possibly hope to represent it in an accurate way? Following this line of thought, the practicality of trying to represent uncertainty must be questioned. After all, if the representation chosen and adopted is merely an approximation of the uncertainty which is actually apparent, does this representation of the uncertainty introduce a higher level of uncertainty into the problem than that which would be apparent if uncertainty had not been used? This argument, associated with the point of view that any true "expert" system should intrinsically incorporate the ability to reason with uncertainty in its rules, seems to indicate that in many cases the introduction of specific uncertainty representations is both pointless and damaging. The uncertainty attached to the chosen representation of the uncertainty will breed further uncertainty, until the level of certainty which can actually be attached to the solution which has been given is minute or meaningless.

How Subjective is Uncertainty?

It must be recognised that a certain level of subjectivity is associated with the uncertainty representations which are adopted in any KBS. Unfortunately, as has already been stated, the understanding of the uncertainty which is apparent and how we tend to deal with it is generally low, and it is often left to an individual (or at best a group of individuals) to interpret the uncertainty which is apparent and represent it in some way. An additional problem is the intended users' interpretation of the uncertainty representation which has been adopted by the KBS developer. This interpretation will again introduce a certain level of subjectivity into the diagnosis.

Fuzzy reasoning is a prime example of an area where subjectivity can greatly influence the interpretation of the uncertainty which is present in a situation. Take the expression:

$$Tall(Peter) = 0.75$$

This is a standard fuzzy reasoning expression, as will be shown by the following section on fuzzy reasoning. This expression applies the principle commonly used that 1 means entirely true and 0 represents entirely false. However, unlike other representations of this type, fuzzy reasoning allows in-between expressions to be adopted which represent how true a statement is felt to be.

So, the above expression illustrates that the statement "Peter is tall" is 0.75 true. Now, the interpretation of this statement is dependent on the opinions of two people: the first person is the one who decided what measure was actually considered to be tall and who thus attached a level of 0.75 to Peter, and the second person is the one who has to interpret this rule and who may well have a different perception of what "tall" actually is. One may argue that in this situation standard values for tall should used, for instance over 6 ft for a male could be considered tall and the fractional values could be taken from there. However, this is introducing a level of exactness which may not be available, and which somewhat contradicts the expression, and indeed the benefits of, fuzzy reasoning.

Is it up to the Expert to Reason Uncertainty out of the Domain?

Some would argue that expertise constitutes reasoning with uncertainty, therefore it is unnatural to separate the uncertainty from the reasoning which is taking place. It is therefore the expert's responsibility to reason with the uncertainty which is present and still produce a satisfactory answer.

This argument is put forward most clearly by Chandrasekaran and Tanner (1986). Here, it is stated that people reason with many types of uncertainty in many different ways, and thus it is not sensible to try and associate and represent the uncertainty in a KBS using one normative approach. They suggest that different approaches should be used to represent different types of uncertainty. This inherently brings problems when these uncertainties have to be combined to attain an overall level of uncertainty. Consequently, Chandrasekaran and Tanner (1986) argue that it is more sensible and practical to allow the systems to use the heuristic rules developed by the experts themselves to deal with uncertainty, so that the systems, by dealing with uncertainty in the same way as the experts would, better emulate their behaviour.

We sympathise with this point of view: it does seem strange that people try to attach numerical measures of uncertainty when it is inherently a non-numerical problem in everyday life. However, as the remainder of this book will show, we are not convinced that the best role for KBS lies in emulating human expert behaviour. We feel that, in some cases, the

KBS could be used to supplement and even replace some of the uncertainty which is apparent in expertise. Therefore, the system would behave more accurately and more reliably than the experts, enabling the KBS to "out-perform" the existing expertise.

Uncertainty Management – An Overview

There are three main aspects to the uncertainty problem (Bhatnagar and Kanal, 1986). These are:

- Representation of uncertain information.
- Combination of bodies of uncertain information.
- Drawing of inferences using uncertain information.

All of these issues bring their associated problems, which are discussed in the following sections. The first consideration is concerned with how the uncertainty can be represented. A number of techniques are currently available which all claim to be better than any of the alternatives. The main ones are discussed in the following sections.

The latter two issues can be considered to be concerned with the propagation of the uncertainty. This is generally recognised as being the more problematic area of uncertainty management.

These two sections will be dealt with together as they essentially involve the same considerations, and the problems associated with the two issues tend to be dependent.

All uncertainty representation techniques individually exhibit advantages and disadvantages, and when used in theory they do not prove to be too problematic. However, it is when these uncertainty measures have to be propagated that the significant problems arise. Dealing with single evidences is relatively straightforward; but when several pieces of evidence need to be combined, difficulties become apparent.

It is recognised that the choice of model which is used for assessing and combining probabilities or uncertainties is more important than how the uncertainty is reflected in the syntax or structure of the KBS (Cohen, 1985; Dym and Levitt, 1991).

Combining or aggregating information and its associated uncertainty is a crucial part of reasoning with uncertainty. This is because the only formal, logically correct approaches available for combining uncertainty measures assume that the probabilities being aggregated have certain attributes, such as statistical independence, which may, in practice, be difficult or impossible to ascertain. Furthermore, without the mathematical precision which is intrinsic to the theory of probability, contradictory or exaggerated confidence levels may be generated.

Many propagation formulae have been put forward and care must be taken to choose an approach which is both accurate and effective, as many of the propagation techniques which have been suggested are no better than the commonly criticised assumption of total independence.

Therefore, in the following sections, the difficulties of propagation are discussed together with the uncertainty representation techniques.

Detailed examples of these propagation techniques are given in Castillo and Alvarez (1991).

Representation and Propagation of Uncertain Information

On first appearance, trying to represent uncertain information appears to be an ambiguous area: after all, as discussed earlier, any representation of an uncertain situation can at best only be an approximation, hence introducing a certain level of uncertainty in itself.

There are fundamentally two approaches to representing uncertainty: numerically and non-numerically, Each of these two approaches incorporates a number of different techniques which are described briefly below. As has already been stated, detailed descriptions of these approaches will not be included here as many are too complex to be covered fully in the limited space which is available. Only an overview is given.

Numerical Approaches
Probability

The oldest measure of uncertainty and the most intuitive is probability. Probability is a mathematical theory which provides a means of dealing with knowledge about truly random events, and it is the only representation of uncertainty which is used in everyday language (i.e. in bookmakers' odds, chance of rain etc.). Probabilities can be converted to odds in favour quite simply. Taking O to represent the odds in favour and P to represent the probability:

$$O + P/(1 - P); P + O/(1 - O)$$

Bookmakers' odds are generally quoted as odds against. This can be converted quite simple by:

$$O = 1/A$$

where A is odds against.

Despite the advance of more "sophisticated" (and certainly more complex) techniques for representing uncertainty, many researchers ardently defend the use of probability in the development of KBS (Cheeseman, 1985; Lindley, 1987), not least because it is the only measure of certainty which has comparatively firm theoretical foundations. Nevertheless, probability is still criticised (Shafer 1982; Zadeh 1983); primarily for the high number of parameters involved and for the difficulties in estimating accurate probabilities from data. Most of the alternative ways of numerically representing uncertainty rely on a probabilistic structure, however, and it is arguable whether they overcome the difficulties experienced when using straightforward probabilities. This issue is discussed in the following sections.

Part of the problem of representing uncertainty lies in the fact that there are two approaches to the mathematical definition of probability (Parsaye and Chignell, 1988). Classically, probabilities are assessed on the frequency of occurrence of an outcome in a series of identical experiments or trials, the standard example of which is tossing a coin. These are generally known as objective probabilities. The alternative definition, which is the Bayesian definition, states that probability should reflect the degree of belief that one has in a fact, outcome or hypothesis. These are generally known as subjective probabilities. For a more detailed description of this distinguishment see Buchanan et al. (1985).

This is an important distinction is KBS development because KBS are generally built to express expertise in domains where repetitive experiments are not possible (Dym and Levitt, 1991). Therefore, degrees of belief seem more feasible than actual outcomes, and thus it is apparent that Bayesian theory is more appropriate for incorporation in KBS than the alternative definition of probability. Therefore, this theorem will be concentrated on here.

Bayesian Probabilities

Bayes' theorem enables the computation of relative likelihoods between calculating hypotheses on the strength of available evidence. It relies on the formula:

$$LR(H : E) = P(E : H)/p(E : H')$$

where LR is the likelihood ratio, E is the probability of the event or evidence given a particular hypothesis H. LR is obtained by dividing this probability by the probability of the evidence given the falseness of the hypothesis H'.

The likelihood ratio can be used to adjust the odds in favour of the hypothesis in question if it becomes known that the event E has occurred. Likelihood ratios must always be positive and similarly values of zero and infinity are conceivable in theory, but in practice should be avoided whenever possible.

The entire Bayesian updating scheme can be encapsulated in the expression:

$$O'(H) = O(H) \times LR(H : E)$$

where $O(H)$ is the prior odds in favour of H and $O'(H)$ is the resulting posterior odds given event E as determined by the likelihood ratio.

Information from a variety of knowledge sources can then be combined by simple multiplication. The likelihood ratios can be obtained from a simple two-dimensional frequency table showing how often each recent event occurs under each hypothesis.

To summarise, the likelihood ratios used in Bayes' theorem have two great advantages:

1. They allow the combination of several independent sources of evidence.
2. They can be easily adjusted if the evidence itself is uncertain.

PROSPECTOR (Duda *et al.*, 1979) is an example of an expert system which used Bayesian theory to cope with the uncertainty which was apparent in the domain. It uses Bayes' theorem specifically to pull together information which is available from disparate sources, making use of the property that Bayes' theorem provides for computation of relative likelihoods between competing hypotheses. Thus information from different sources can be combined.

The problem of combining probabilities can be reduced to the calculation of probabilities which are conditioned by all pieces of information. However, as already mentioned, this combination relies on the assumption of independence: the validity of which is questionable in many "real-life" situations. There is almost inevitably going to be some level of dependence in any real set. By assuming independence, it is possible to obtain completely erroneous results. Instead, it may be preferable to minimise the dependence and work with a relevant dependence model which includes the applicable information so that an appropriate result can be extracted.

Because of these problems, most Bayesian reasoning programs incorporate checking routines which search for duplicate rules or rules which overlap.

Certainty Factors

An alternative way of measuring uncertainty is by using certainty factors. These were first developed and used for the well-known MYCIN expert system (Shortliffe, 1976). This scheme of measurement assesses the confidence that can be placed on any given conclusion as a result of the evidence so far. An example of a rule in MYCIN is:

MYCIN
RULE027
If: 1)   The site of the culture is blood, and
*   2)   The organism was able to grow aerobically, and*
*   3)   The organism was able to grow anaerobically*
*   Then: There is evidence that the aerobicity of the organism*
*   is facultative (.8)*
*   or anaerobic (.2)*

where .8 and .2 are the associated certainty factors.

Many see certainty factors as a way of overcoming the problems of weighing up disparate sources of information. Others argue, however, that although in theory this is true, the practicality of certainty factors is questionable.

Certainty factors (CF) involve two other measurements: the measure of belief (MB) and the measure of disbelief (MD). The certainty factor is then taken to be the difference between these two component measures (Forsyth, 1984):

$$CF[h:e] = MB[h:e] - MD[h:e]$$

Therefore, $CF[h:e]$ is the certainty of the hypothesis h given evidence e.

Similarly, MB and MD are the measures of belief and disbelief respectively of h given e.

CFs can range from completely false (−1) to completely true (+1), with fractional values in-between representing intermediate levels of certainty. Zero is taken to be mean ignorance. MDs and MBs range from 0 to 1, so the CF is essentially a simple balance between for and against.

It is interesting to note that this formula does not take into account conflict of evidence (i.e. that the MB and MD are both high) as opposed to the lack of evidence (MB and MD both low) which could be important.

None of the measures (that is, neither CF, MB nor MD) are probabilities. Although they obey some basic rules of probability, they are both statistically based. This does not apparently offer any real advantage over fuzzy logic. However, Shortliffe's uncertainty factors in MYCIN also include an updating formula which allows new information to be combined to produce up-to-date results. This applies to the measure of belief and disbelief associated with each proposition. The formula for MBs is:

$$MB[h : e_1, e_2] = MB[h : e_1] + MB[h : e_2] \times (1 - MB[h : e_1])$$

MDs are updated in the same way. Therefore the effect of the second piece of evidence (e_2) on the hypothesis h given earlier evidence e_1 is to move a fraction nearer to the certainty as indicated by the second piece of evidence.

The two important factors of this formula are:
1. The order of e_1 and e_2 does not matter.
2. MB and MD move asymptotically towards certainty as supporting evidence builds up.

This certainty factor scheme also allows for the possibility that the inference rules and the data may both be uncertain. Each rule has an attenuation (a number from 0 to 1) which indicates its credibility.

Shortliffe does not attempt a theoretical justification for these methods and indeed many question the theoretical accuracy of such measures (Forsyth, 1984; Bhatnagar and Kanal, 1986; Horvitz and Heckerman, 1986). However, MYCIN has proved to be successful (as has EMYCIN and its successors), and so the fact that this well known system has been successful must indicate that the methods must be worth consideration and also further investigation.

Arguments have been put forward, however, (Horvitz and Heckerman, 1986), that these certainty factors are misrepresentative because they measure the current level of certainty as opposed to the change in certainty. Horvitz and Heckerman (1986) claim that there is an important distinction to be drawn between the two, as one represents an updated belief whereas the other (if a distinct change in belief takes place) represents an absolute belief. They contend that measures of belief or belief update should be combined in a way which is consistent with the expertise they are associated with. This would encourage fewer errors.

The propagation of the measures of the belief MB and the disbelief MD of the certainty factors CF is usually carried out by using the following formulae:

$$MB(h,e_1 \cap e_2) = \left\{ \begin{array}{l} 0 \text{ if } MD(h,e_1 \cap e_2) = 1 \\ MB(h,e_1) + MB(h,e_2) - MB(h,e_1) \times MB(h,e_2) \\ 1 \text{ if } MB(h,e_1) = 1 \text{ or } MB(h,e_2) = 1 \end{array} \right\}$$

$$MD(h,e_1 \cap e_2) = \left\{ \begin{array}{l} 0 \text{ if } MB(h,e_1 \cap e_2) = 1 \\ MD(h,e_1) + MD(h,e_2) - MD(h,e_1) \times MD(h,e_2) \\ 1 \text{ if } MD(h,e_1) = 1 \text{ or } MD(h,e_2) = 1 \end{array} \right\}$$

$$MB(h_1 \cap h_2,e) = \min(MB(h_1,e), MB(h_2,e))$$
$$MD(h_1 \cap h_2,e) = \max(MD(h_1,e), MD(h_2,e))$$
$$MB(h_1 \cup h_2,e) = \max(MB(h_1,e), MB(h_2,e))$$
$$MD(h_1 \cup h_2,e) = \min(MD(h_1,e), MD(h_2,e))$$

It should be noted that these formulae are inexact. The first two refer to the combination of evidence under the same assumption and the last four refer to the combination of the two hypothesis under the same evidence e.

However, in practice it has been shown that the reliability of these theorems is very questionable, and therefore the user must be warned to use them with care (see Castillo and Alvarez, 1991).

The Dempster–Shafer Theory of Evidence

The fundamental idea behind the Dempster–Shafer (Shafer, 1976) theory relates to the fact that often we do not know individual prior probabilities, therefore any assignment of a probability value is arbitrary. However, often we do know something about a group of objects or events, so that we can assess a strength of belief for that class. Dempster–Shafer offers a calculus which assigns and manipulates weights to sets and subsets as opposed to individual items as a reasoning process goes along, thus being more dynamic than many approaches to uncertainty representation.

It can now be seen that the Dempster–Shafer theory, like Bayesian theory, relies on degrees of belief to represent uncertainty. However, unlike Bayes' theory it allows the assignation of degrees of belief to sub-sets of hypotheses (Cohen, 1985). In addition, in this approach (see Shafer, 1976), a distinction is made between uncertainty and ignorance. Belief functions are specified instead of probabilities which enables bounds to be put on the assignment of probabilities, so that the probabilities do not have to be specified exactly. The method also provides mechanisms for combining evidence and producing the appropriate belief functions. Therefore, essentially, when the bounds determine the probabilities exactly, this approach reduces to the probability theory.

Because the Dempster–Shafer theory incorporates probability as a special case, it inherits many of the problems associated with it. It also has the disadvantage that it is more complex and therefore more computationally inefficient. However, the Dempster–Shafer theory is important as it illustrates the effect of ignorance on reasoning with uncertain knowledge.

Dempster (1968) proposes the following formula for combining two units of evidence m^1 and m^2:

$$M(A) = \left\{ 2k \sum_{B} \sum_{C=A} M_1(B)M_2(C) \quad \begin{array}{l} \text{if } A \neq 0 \\ \text{if } A = 0 \end{array} \right\}$$

K is the normalising constant that makes m a probability (Chatalic et al., 1987). However, like the certainty factors, these theorems prove to be less than reliable when used in real-life examples. Many would seriously question Dempster's Propagation formula (Castillo and Alvarez, 1991a) and so again these formulae should be used with great care, if at all.

Fuzzy Logic

Fuzzy logic (also called possibilistic logic) was invented by Zadeh (1965). He extended classical Boolean logic to real numbers. As with Boolean logic, in fuzzy logic, 1 represents truth and 0 falsity. However, in fuzzy logic, the fractions between 0 and 1 indicate partial truth. The aim of fuzzy logic is to allow the subjective quantification of the imprecision or vagueness which we so frequently encounter. Information may be unreliable because we are unable to obtain it with sufficient precision (for instance, when measuring data in practical testing) or because we are unable to define the concept lying behind the information. In the former case, the randomness of the situation is being recognised, that is, the event is clearly defined but the imprecision attached to it is not (Dym and Levitt, 1991). The latter case is an example of where the fuzziness of the description is being recognised (Zadeh, 1975).

A common example of fuzziness is:

$$P(\text{tall}(X)) = 0.75$$

which states that the proposition X is tall is 0.75 true. From this we can gauge that the person is taller than average, which somewhat overcomes the ambiguity of the expression "tall". The same ambiguity can be seen with expressions such as long, thin and old, all of which represent somewhat subjective judgements. Cohen (1985) cites a more interesting example: that is, when is a beer a *cold* beer?

AND, OR and NOT are used in fuzzy logic to combine non-integer truth values:

$$p1 \text{ AND } p2 = \text{MIN}(p1,p2)$$
$$p1 \text{ OR } p2 = \text{MAX}(p1,p2)$$
$$\text{NOT } p1 = 1 - p1$$

This enables pieces of evidence to be combined in a consistent way. Fuzzy logic is used in the decision support system REVEAL (Jones, 1984).

However, fuzzy logic has its weaknesses and many researchers (particularly supporters of probability) protest that it does not exhibit any advantages over ordinary probability: in fact, some claim that it does not represent uncertainty as accurately. After all, if a person is 35, how true is that statement that they are old? Is it 0.5 because 35 is approximately half a lifetime? Or would 0.4 or 0.6 be more accurate? Somehow, a mapping

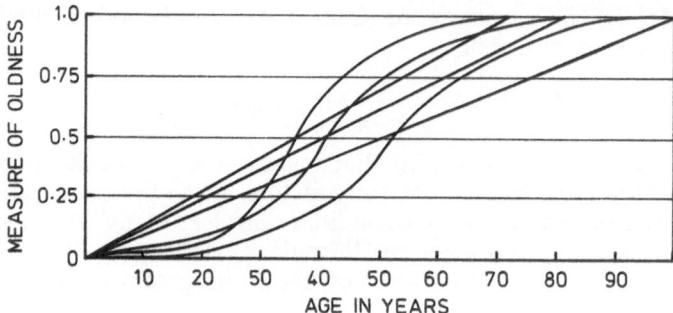

Fig. 3.1 Alternative mappings of age in years to a measure of oldness – any one of which can be considered correct.

function which describes the relationship must be decided (Fig. 3.1) (Forsyth, 1984). However, there are many different mapping functions which could be used to describe this relationship, and therefore the choice of function will be somewhat arbitrary (as discussed earlier, this introduces uncertainty in itself). One way in which this issue is side-stepped in KBS development is to allow the user easily to modify the mapping functions involved, hence discovering whether the precise shape of the relationship is critical or not (Forsyth, 1984). This approach has been adopted in RE-VEAL (Jones, 1984, 1989).

Possibility or fuzzy theory focuses not only on how the possibility distributions are obtained but also on how the numerical calculations which obtain the fuzzy variable might be done. Some examples of the predicates which may be used in rule-based systems are given in Buchanan (1985). The fuzzy reasoning approach exhibits a great deal of potential, but as yet there have not been many applications of it in practical KBS development. A discussion of potential engineering applications can be found in Chang *et al.* (1988).

Combining fuzzy sets is a complex subject which cannot be covered here in the space which is available. The main difference from the previous propagation formulae which have been stated is that the formula which is used for combining fuzzy sets is taken directly from the possibility function axioms whereas the others have been arbitrarily chosen.

A distinct problem with combining fuzzy sets is that the mathematical models used do not reproduce the physical properties of many real cases (Castillo and Alvarez, 1991a), and hence the use of possibility functions is questioned in many situations.

Non-Numerical Approaches

Often, the necessary information which is available is incomplete. As has already been stated, this introduces a level of uncertainty into the problem as, in these cases, somehow the significance of these unknowns has to be assessed. For such cases, symbolic, non-numerical characterisations can be more useful than numerical representations (Doyle, 1983; Dym and Levitt, 1991).

There are four commonly used arguments for using non-numerical or qualitative approaches (Winstanley, 1991):

1. In AI it is often important to capture as many generalisations as possible. Numerical methods tend to be too detailed when compared with other non-numerical approaches which take a more general view.
2. Many aspects of non-monotonic reasoning (e.g. default reasoning) do not fit comfortably with numerical approaches.
3. Commonly there is a lack of quantitative knowledge from which to build a numerical representation, which leads to imprecise figures being used.
4. Numerical representations of uncertainty make systems difficult to expand and maintain because of the interrelation of the numerical factors.

These viewpoints indicate that numerical representations do not adequately encapsulate and represent uncertainty. This is not entirely true: the increasingly popular opinion is that no one approach is adequate for representing all types of uncertainty (Chandrasekaran and Tanner, 1986; Fox, 1986; Nutter, 1987; Pollock, 1987) and that approaches to uncertainty should be chosen carefully and with discretion. However, non-numeric representation techniques do exhibit some distinct advantages over numerical representations, as will be discussed in the following sections.

There are fundamentally two types of non-numeric approach: logic-based approaches and non-formal representations. Some may argue that fuzzy logic is indeed a non-numeric approach, but in this book we have qualified it as a numeric approach as it deals with uncertainty representation in a quantitative manner as opposed to a qualitative way. We have therefore assumed that the way in which fuzzy logic deals with uncertainty fits more comfortably into the numeric approaches section than the non-numeric section.

The other approaches to non-numerical representation are briefly covered in the following sections. For more detailed information on these approaches see Lewis (1973), Rich (1983), Fox (1986), Klahr and Waterman (1986) and Nilsson (1986).

Logic-Based Approaches to Uncertainty Representation

Logic-based approaches to uncertainty representation rely primarily on symbolic techniques. Unlike numerical methods, they concentrate on the relationships which exist between the propositions to reason about uncertainty. PROLOG and other logic-based languages provide a good basis for such uncertainty representations, as the mechanisms within them support the techniques used. In order to enhance the usefulness of standard logic, two approaches are used:

- Using "deviant logics" which incorporate different or more advanced semantics to enable the classical logic to deal with different proof procedures.
- By extending the vocabulary of logic.

When reasoning under uncertainty in logic, generally the approaches to representation operate at a meta-level, the uncertainty being represented by some special addition to the fundamental logic representation (Winstanley, 1991). Some would argue that the uncertainty should not be dealt with at a meta-level but instead it should be dealt with in the same way as the other information about an object; i.e. at an object level. There are obvious benefits to this, but it can make the actual programming complex. This argument is too complex to be dealt with in detail here. For more information see Fox (1986).

Modal logic is one way of extending standard logic so that it can represent the concepts of necessity and possibility. Modal operators can be introduced into the language (e.g. "it is possible that" or "it is necessary that") and can be used to precede a logic statement and thus extend the flexibility of the logic representation. They therefore can provide a good form of uncertainty representation. For more details on this see Winstanley (1991).

Non-Formal Representations of Uncertainty

Many KBS developers are switching to less formal methods for representing uncertainty in domains where uncertainty is an integral part of the knowledge representation. For example, Alvey and Greaves (1990) and Fox (1986) have investigated the use of everyday verbal expressions to represent the level of uncertainty which can be attached to certain statements, such as "very", "not very", "hardly" etc. Alvey and Greaves (1990) feel that by using these informal techniques, they are dealing with uncertainty in "the same way as the expert handles it". They are also of the opinion that by using numerical representations of probability, you run the risk of alienating the experts as they are uncomfortable with the approaches being used.

Many informal approaches may easily be linked with a logic representation, of the type which have been discussed in the previous section. However, they can also be used as separate entities with different programming environments.

Obviously the different ways of representing uncertainty in an informal way are numerous: too many and diverse to be discussed in detail here, as there are no limitations on the expressions which can be used. The expressions are generally chosen to reflect the uncertainty expertise which is being dealt with and cover a spectrum of opinion.

For example, Fox (1986) has investigated the use of terms such as "possible", "probable" and "plausible" together with the negative equivalents ("impossible", "improbable", "implausible"). Although these are recognised statistical terms, as they are usually linked to some point on a probability distribution, Fox (1986) has used them in a less formal way, as he has defined them in logic programming terms. By not associating these terms to numerical values, the logic can be used in a more effective and flexible way.

Informal approaches to uncertainty representation have the advantage that the developer is not attaching numerical judgements to an area which is essentially non-numerical. Also, it must be recognised that these approaches do allow the uncertainty to be represented in a way which is both familiar and appealing to the expert(s) concerned, and which may more fairly reflect the levels of uncertainty involved in the domain (Alvey and Greaves, 1990; Hunter, 1991).

However, using an informal approach has the disadvantage that combining the various uncertainty qualifications can be difficult (especially when two extremes are present, such as "not at all" and "often"), as there are no pre-defined methods of doing this. Also, the interpretation of these informal qualifications tends to be highly subjective, which can prove to be very problematic in establishing a consistent set of results. As Hunter (1991) states, before using any of the non-numeric, informal approaches, the developer needs to analyse:

"... the efficiency of interaction, the minimisation of ambiguity and the accuracy of the computer solution to that required application."

Therefore, non-numeric approaches to uncertainty representation can introduce a high level of inaccuracy and subjectivity, the dangers of which have already been discussed.

Conclusions

The most common forms of uncertainty representation have been put forward in this chapter. However, it must be recognised that this list is not exhaustive.

The subject of uncertainty is both complex and confusing. It is an area which is the centre of attention for much disparate research which is being carried out by many different groups. Consequently, much disagreement about what uncertainty is, how we, as people, deal with it and how best to represent and manipulate it exists.

Without an agreed model of how the human mind processes uncertainty in the real world, inevitably there can be no single method with which this uncertainty can be represented. When focusing on KBS development, the system builder needs to concentrate on the management of uncertainty: trying to identify the cause and effect of the uncertainty which is present in the domain and choosing the approach which is most sympathetic to these criteria. In many cases, (as has been our experience) it may well be found that no formal incorporation of reasoning under uncertainty is required: instead it may seem more feasible and sensible to allow the assessment of the uncertainty to be dealt with within the "expert" rules. Trying to be precise about uncertainty is inevitably a dangerous route to follow as it is inherently an imprecise subject. Therefore, the problems which can be encountered when trying to represent and propagate uncertainty should be borne in mind.

All the approaches which have been discussed in this chapter have their supporters and their critics. Fundamentally, it is apparent from the mass of literature which is available on the subject that much of the research into uncertainty representation has been carried out on a purely theoretical basis without regard for the practical implications of the representations and propagation techniques which are proposed. Although certainty factors, Bayes' theorem and fuzzy logic have all been used in applications, the amount of research which has concentrated on developing these techniques in a practical way seems limited compared to the enormous amount of investigation which is carried out into comparing the various techniques and contesting existing theories.

Overall, the main conclusion which can be drawn is that only the developer can decide whether uncertainty is relevant for the KBS being developed and if so, which type of representation and propagation techniques are applicable. As has already been mentioned, to date we have not found the inclusion of uncertainty representation to be of any real benefit to the systems which we have developed: it is apparently not formally manipulated by the experts which we have dealt with and indeed they showed a perceptible level of discomfort when the idea of uncertainly measurements was put forward. Also, we felt that the level of certainty which could be attached to the uncertainty was low and therefore the worth of uncertainty measurements had to be questioned. We feel that this bias is inevitably reflected in the discussion in this chapter. However, we do feel that there is scope for replacing human uncertainty (that is, uncertainty which is introduced owing to the heuristics which are used) in KBS to create improved systems which "out perform" experts in certain areas.

Chapter 4

Design

Objectives

This chapter aims to:

- Give an overview of design and to describe briefly the different approaches which are used for design.
- Look at the way in which design is carried out in practice.
- Investigate two ways of interpreting the design method.
- Discuss the possible benefits which KBS may bring to design.

Introduction

Chapter Contents

This chapter is intended to outline what design consists of and place design activity in its correct historical perspective. The different approaches to design by discipline, nationality, product and client/purchaser are contrasted. The chapter also includes details of actual studies of designers at work to show what design really is, and covers so-called design science or how people think that design should occur. The former is of especial interest as it gives further clues as to what constitutes expertise in design and the cognitive process that occur.

Finally we look at how KBS can add to the design process without detracting from the designer's contribution.

An Overview of Design

Theodore von Karman defined the difference between engineering and science in the following way:

"The scientist explores what exists; the engineer creates what isn't."

In order to create any reasonably complicated and planned artefact it is necessary to produce a design which then becomes the plan of what is to be made.

As with all generalisations, there are exception to von Karman's definition. For example, does this mean that architects are engineers or should a special category be added to the definition for them? Possibly it would be better to replace engineer with technologist, but even then the architects tend to prefer to think of themselves as artists rather than technologists.

Certainly in the English-speaking world a myth persists with the general public that to be creative one has to be an artist, hence presumably the architects' affinity for art rather than technology. This is an odd attitude for a culture which played a major part in developing the modern advanced society which is based on the creation of products by technologists.

Jones (1980) presents an interesting view on the design process in which he states

". . . designing should not be confused with art, with science or with mathematics. It is a hybrid activity which depends for its successful conclusion upon a proper blending of all three."

Jones then goes on to discuss the difference between the scientist, the artist and the designer. The former tries to understand and explain the things around him/her. The artist manipulates that which exists, albeit paint, stone or some other medium. Use is made of imagination during the work, and sketches and plans are employed, but it is accepted that much of the work is impulsive.

The designer, however, has to conceive of that which is to exist. He/she does not work with the actual materials of whatever is to be built and usually cannot substantially amend the design during the manufacturing/ construction process, but instead has to think through all aspects of the product before it is created.

There are other aspects of design which need to be born in mind while reading this chapter. As Stauffer *et al.* (1987) point out, in this century manufacturing production has increased many times over but the productivity of the designer has only improved very slightly. Also as Gill (1987), quoting the Corfield report, states:

"The main elements of cost and desirability are determined not on the shop floor but in the design. Poor design wastes valuable labour and materials".

Design is one of the most important and challenging activities that mankind undertakes. Our ability to create new artefacts has enabled us greatly to extend our range of achievements and vastly to improve our lifestyle. It also enables us to exert some degree of control over our environment. These capabilities form the basis of what we call civilisation. To accomplish this level of performance would be impossible without being able to conceive and plan things. This latter process is what we generally call design. From such things as a humble cheese grater through to such complex objects as robots with their associated software, little can be accomplished without first going through the design process. Oddly enough, although in most disciplines design has enabled all these technological advances to be realised, the design process itself is largely unaffected by technological advance. The processes used, with the exception of some help from computers in analysis and drafting, are similar to those which were in use a hundred years ago.

Definitions of Design

When one starts to search for definitions of design in the existing literature, the number and diversity of views are overwhelming, and yet in general discussions people seem to use the terminology of design in similar ways. This paradox possibly arises because of the personal nature of the design process, but this can only be a partial explanation. Susan Finger (at the time the Director of the Design Theory and Methodology programme for the USA's National Science Foundation) asked the question, what constitutes design? The replies received were (Talukdar *et al.*, 1988):

- Design is satisfying constraints and meeting objectives.
- Design is problem solving.
- Design is decision making.
- Design is reasoning under uncertainty.
- Design is search.
- Design is planning.
- Design is an interactive process.
- Design is a parallel process.
- Design is an evolutionary process.
- Design is a mapping from a functional space to a physical space.
- Design is like a game.
- Design is creative and is inexplicable.

Although some of the more obscure definitions can be discounted, there is in the majority of the above an element of what most people think design consists of. Yet none of the above fully describes design.

Partridge (1981) describes design as being a "wicked problem" in so far as design has no definitive formulation, no stopping rule, no solution in the absolute sense and each problem is unique.

One of the classic texts on design (Pahl and Beitz, 1988) describes designing as "... the intellectual attempt to meet certain demands in the best possible way". This is a more complete definition than any of the above, but it fails to include the fact that design is an activity which produces a description of something which is yet to be created, indeed a description which enables that artefact to be created.

There are many other attempts to describe what design consists of (see, for example, Kalay, 1987 and Mostow, 1985), but it can be concluded that as yet there is no single accepted definition. Finger and Dixon (1989a), in a discussion on design research, highlight the problem. Although people undertake design, our understanding of the design process is as yet incomplete. Without a good level of understanding, a robust definition obviously is not possible.

However, as we shall attempt to show in the following parts of this chapter, there is much which has been discovered about design and which is pertinent to the development of design KBS.

Historical and Sociological Perspectives

History tells us that the first artefacts created by humans were probably stone tools. Even at this level some degree of design is needed in the specification of the stone type and the shape and sharpness of the tool. However, at this level the design process is simple enough to be accomplished without the need to resort to some form of documentation. Indeed, at the craft level many things are built without the need to resort overtly to what we generally think of as design. Also, in medieval times, very large complex buildings were created using craft-based techniques. So one can conclude that, within constraints, much has been and continues to be achieved using this method.

However, several factors can intervene which may create the need for the design to be recorded in some form which is comprehensible to others. These factors include:

- The requirement for mass production.
- The design of a complex object.
- The involvement of more than one person in the design/create process.
- The need to have the design checked.
- The obtaining of competitive tenders for realising the designed object.

Influences such as these, which have their origins in architecture but came to the fore predominantly during the development of industrialised societies during the 19th century, have caused the design process to develop to the point where it can involve large teams of people. Not all of these are necessarily technologists, some may be lawyers and accountants, for example.

Thus the development of design came about as a response to the increasing complexity of products, and manufacturing and construction processes.

Rather than being systematically developed, the processes used for design have evolved. They are treated by many as a skill to be acquired largely by practice on the job rather than by formal teaching. The way in which design knowledge is acquired is generally assumed to have a profound influence on the designer's later attitudes to the design process (Krieger, 1987), and yet the development of design skills (especially for conceptual design) in a person tends to be achieved in a fairly *ad hoc* manner.

In the English-speaking world, in educational establishments which teach engineering, what is generally taught as design is in fact analysis (for a good example, see Shigley, 1972). This also spills over into research with design research being largely concentrated on analysis (usually the development of numerical techniques). Thus when the graduates of these institutions go into industry, their appreciation of design tends to be that it is highly mathematical and consists of analysis. Their knowledge of the other aspects of design, and particularly conceptual design, is woefully inadequate and it is only through work experience that this extra knowledge is acquired.

The available evidence suggests that, in German-speaking countries, design is taught using a much more systematic approach (see, for example, Hubka, 1982, Eder, 1987, Hubka, 1987, Pahl and Beitz, 1988). Therefore graduates of these countries have a much better appreciation of what constitutes design and will presumably follow a more structured approach to the design process, although there is no evidence in the literature to say whether or not this is the case.

The above discussion on approaches to design is of importance from a number of aspects. Firstly, as the pace of technological change increases, designers are being asked to make decisions about products which utilise techniques and materials about which relatively little is know, certainly with respect to their long-term behaviour. Thus the job of the designer is becoming harder. To help with this problem it is essential that structured, quality-assured approaches to the design process are followed. If the whole ethos of conceptual design is that it is an unstructured skill which is only acquired by experience, then this may well be difficult to achieve.

Also, in the English-speaking world it is unlikely that major improvements in design education will occur. Generally, social systems contain far too much inertia for such developments, and so it is vital that different ways are found to introduce structure into the design process. Well developed computer systems based on KBS offer potential solutions to some of these problems.

Components of Design

Classes of Design

Before describing the individual components of the design process it is necessary to identify what broad classes of design activity exist. Inevitably, when one tries to place any human activity into classes, then there are areas on the boundaries between classes which do not fit conveniently into a slot. However, given this proviso, design can be split into three classes (Brown and Chandrasekaran, 1985):

- Class 1 – Major inventions or completely new products.
- Class 2 – Designs which involve substantial innovation.
- Class 3 – Routine design which involves selecting among previously known, well understood sets of design alternatives. Although at each stage of the design the choices may be relatively simple, the task is still too complex for it to be achieved by copying a previous design.

Pahl and Beitz (1988) give three similar categories which they call original design, adaptive design and variant design. They also quote figures collected by the VDMA (German Association of Mechanical Engineering Companies) for 1973 which show that, of the total design effort, 55% was devoted to adaptive design, 25% to original designs and 20% to variant design.

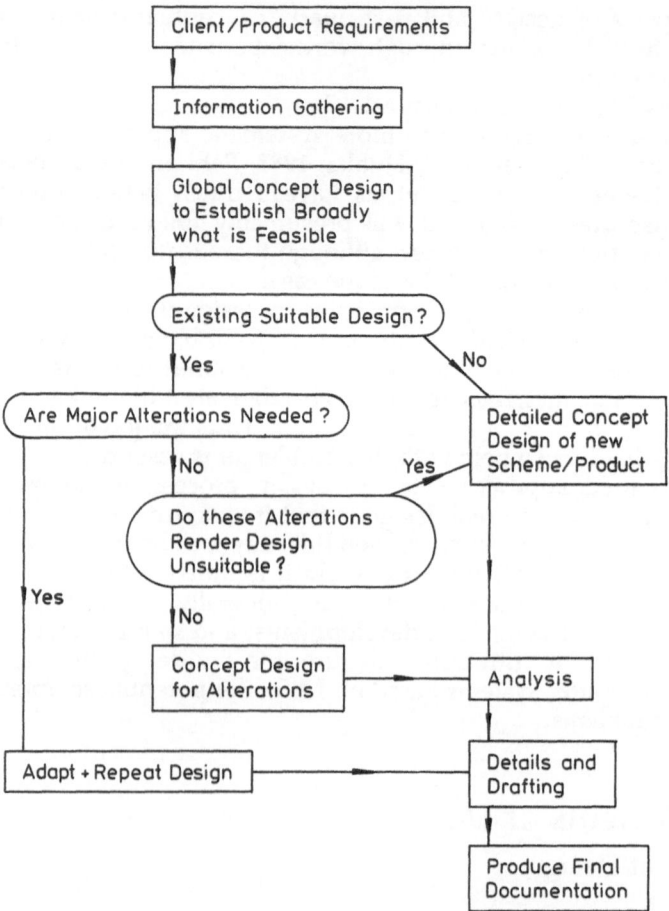

Fig. 4.1 Flow diagram of the design process

Implicit in the description of class 3 type design is the assumption that merely reworking an existing design does not really constitute design in itself unless substantial alterations are made. However, if the designer is undertaking his/her task conscientiously then this is not true because the use of a previous design should be preceded by an evaluation of that design and other alternatives (see Fig. 4.1), and this activity, as will be shown below, is the basis of conceptual design. It has to be admitted, though, that owing to temporal and economic pressures, in practice an existing design is often used without a full evaluation of alternatives.

Subsections of Design

A coarse breakdown of the design process is given in Fig. 4.1, in which it can be seen that initially there is some sort of requirement for a product.

The source of the need may be either an internal requirement of the organisation or there may be an external client of some sort.

The definition of the need is followed by a period of information gathering. For example, for a Civil Engineering scheme this may be a site survey and geotechnical investigation. Also at this stage it is normal to collect details of previous products which have similar characteristics to what is proposed.

There then follows what is described in Fig. 4.1 as "Global Concept Design". This is where, in broad outline the information is evaluated and options are identified. Comparison of the options then takes place. Often, depending on the type of organisation involved, this comparison is undertaken very quickly using largely heuristic knowledge. If the design is undertaken by one person rather than a team, then it may not even be readily apparent that this stage has occurred and in some cases there is doubt that it really does take place to a sufficient level.

If an existing design is found to be suitable then the level of modification which is necessary has to be identified. If this is small then effectively very little further design takes place other than the production of suitable drawings and documents. For all other cases of design it is necessary to prolong the period of conceptual design, because either an existing design is to be modified or a completely new design has to be created.

This further period of conceptual design ideally consists of further consideration of options in greater detail than in the global concept design, with the options being evaluated and gradually reduced in number until the most suitable solution is determined.

Figure 4.1 then shows a discrete step between conceptual design and analysis. In practice, the step is a gradual one with the designer moving from searching for solutions/making judgements to fixing accurately the sizes/capacities of components. Typically, analysis contains a lot of deterministic procedures based on declarative rather than procedural knowledge, although as Burgoyne and Jayasinghe (1991) indicate, there is still a substantial amount of heuristic knowledge used in analysis.

Once again the boundary between analysis and detailing/drafting is not a discrete step as shown in Fig. 4.1. Drafting in the sense that it is discussed here means the final drawings rather than the working drawings and sketches which almost all designers use throughout the design process to aid their thought processes and to provide an memory prompt (see Ullman et al., 1987) for a description of the importance of working drawings and sketches). Finally, at the end comes the boring but essential process of documentation. As well as describing the design in graphical form, it is necessary to provide instructions/specifications and other documents.

As discussed above, the design process is not fully understood and there are as a result of this many variations on the above model. The terminology is confusing because some people tend to call conceptual design "preliminary design", whereas for some areas of activity it is called "strategic planning". However, the level and type of mental activity are the same.

The Effect of Profession and Organisation

Types of Profession

Design is an activity which is undertaken in many different spheres of activity, from the person who plans a garden to an engineering team designing a complex machine. In both cases it is necessary to envisage that which is to be created, although one could argue that in the latter case the consequences of failure will probably be more severe.

The range of professions which undertake design is quite large and includes:

Architects
Chemical Engineers
Civil Engineers
Electrical Engineers
Electronics Engineers
Fashion (i.e. clothes) Designers
Furniture Designers
Mechanical Engineers
Landscape Gardeners
Software Engineers

As discussed above, architects (and fashion and furniture designers) consider themselves to be artists rather than technologists, although they do provide functional solutions to satisfy human needs. However, architects tend to place a lot of emphasis on the appearance of their product; for example, the classic (but undoubtedly apocryphal) anecdote regarding the design of Sydney opera house in which it is stated that the architect just said what the exterior of the building should look like with no concern for how the internal space should be broken up or how the building would be constructed. Like all such anecdotes, there is a germ of truth in what is said in that architects do spend a lot of time considering the external appearance of their buildings or spaces. This, however, is because buildings are prominent objects within the built environment and hence their appearance has a major impact.

Civil engineers (we include structural engineers in this category) tend to think of themselves as constantly designing prototypes in so far as very few of their designs are mass produced, and even in cases where designs are repeated (say, for example, bridges on a particular highway) each case has some unique feature which requires alterations to the standard design. Yet often designs in civil engineering have strong similarities to previous designs, so that although each design is a prototype, often the difference from previous designs tends to be relatively small and therefore there is a gradual evolutionary process from one design to the next, with improvements being made slowly but continually.

In contrast, the general perception of electrical, electronic and mechanical engineers is that much of their design work is for mass production. Just how true this is cannot easily be determined. Certainly many products in these disciplines are mass produced, but it is conceivable that the perception of design activity is biased because of the inevitable preponderance of mass-produced articles compared to prototypes. Possibly, then, a significant amount of design activity in this area is of prototypes, but again, as for civil engineering, many of these will be adaptations or variants of previous designs.

Software engineers seem at first to be a different category and indeed their work does appear to have distinct problems, particularly in the aspects of validation verification and working with large teams. We have no direct experience of the software industry, but our research indicates that possibly of all the above disciplines, the one which creates more prototypes than any other design activity is software engineering, although of course many products are also for the mass market.

Organisational Structures

In this section the influence on the design process of the interactions between the various parties who participate in creating the demand for the product and who will purchase the final article are examined.

First let us examine who decides that a product is needed.

For a product which is to be designed within a company typically the stimulus comes from either:

- Management of some form, be it company or government;
- Technical research which throws up a new idea; or
- Marketing.

Marketing tends to react to what the public already are aware of and so rarely do innovative products come from this area (Clausing, 1985). Likewise, management (certainly in the English-speaking world) tends to consist of lawyers and accountants who are not known for their innovative skills, so any truly new ideas are usually dependent on the technical research departments.

If the product is commissioned by an external client, then the nature of the requirement is dictated by the client but the form is usually determined by the designer. Much of the work of architects and civil engineers is via this sort of mechanism, although the amount of "in-house" design and build developments is increasing.

Whoever constructs the product also has an effect on the design process. If the designer and manufacturer work for the same organisation then it is likely that good links exists between these two factions, and the designer will therefore be well aware of how to design to minimise manufacturing costs. Where the designer and constructor work for separate organisations then often there is no feedback from the constructor to the designer, indeed the designer in some cases has little to do with the final product. This is obviously not a good organisational structure so far as the development of the design process is concerned.

The cost, size, potential risks and sales of a product also have an influence on design. Small, relatively cheap products can be developed by an iterative process of design and test until the major problems have been identified and dealt with. Where products are to be mass produced the payback from minor reductions in the cost of each item can justify the extra design effort whereas if something is to be a "one-off" or only a small production run is planned then a highly refined design is uneconomic. Also, in the aerospace industry although costs are high, the payback from reducing the weight of components and the penalties of failure are such that design and test procedures are used extensively. The other major area where design and test are used is where the understanding of or confidence in the technology is low.

When the cost of the product is high and especially if the product construction times are relatively long and/or the size of the product is large (e.g. a dam), then design and test is obviously not feasible. Models can be used to provide some back up for the designer, but generally the designer is faced with having to ensure that things work correctly without the facility of building and testing prototypes. This tends to make the design more conservative and reduces the likelihood of technological advance.

The means by which the final product is purchased or paid for and the general perception of the product, are major factors in design. The interest of architects in the outer shape of buildings at the occasional expense of functionality has already been discussed. Another good example of products where appearance is all important is the high-performance car. If a sports car is to be successfully marketed it has to look powerful, sleek and fast. It helps if the performance matches the appearance, but this is not essential. As anyone who has driven such a car will attest, the designer usually places much more emphasis on appearance than functionality, with space inside often being at a premium and ingress and egress being only for the more supple members of society. But, then, such cars are wasted on those of advancing years.

If the public perceives an item as being a luxury or a high-quality product, then again external appearance has a major influence on the design, although the functionality may also be of importance. Luxury cars, for example, have to look large and expensive, but in this case their owners require more interior comfort than one finds in a sports car. There are hybrids such as the Jaguar luxury sports cars, but these inevitably involve compromises.

Generally, if something is to be sold through the retail market then its outer packaging and appearance are of importance and the need to offer the salesman "features" to help push sales is vital. Many people have bought video machines which have a multiplicity of features. At the time of purchase it seemed worthwhile paying the extra money for these gimmicks, but research has shown that most of them are never used. Indeed, the only people who seem able to master these complex machines are young children.

If a product is not sold via a retail outlet or if it is sold more on the basis of cost or performance, then the functionality and the price of the equipment becomes far more important. Likewise, if a client commissions a piece of work then, except in the case of prestige buildings or structures, cost and utility tend to dominate the design process.

So all of the above factors can potentially influence the emphasis that is placed on various parts of the design process, but do they fundamentally alter the basic mental process? For instance, are the cognitive mechanisms used by an aircraft designer substantially different from those used by an architect? Our guess is that all designers basically use similar processes.

Design: The Reality

Introduction

There are two quite distinct strands of thought about design. On the one hand there has been a substantial effort by a few people to try and understand how people actually do design, and then there has been the development of processes which try to provide a correct way to design (sometimes called systematic design or design science). A good example of the former is the work of Akin (1986) and of the latter the contribution of Pahl and Beitz (1988).

In this section the work of those who have studied designers at work will be covered. Systematic design will be covered in the following section. The contrast between the two approaches is useful because one describes what actually happens and the other gives guide-lines as to what should happen. However, it is suspected that even when the systematic approach is used for a project, many of the phenomena which are described in this section still occur, albeit within a structured framework (Ullman et al., 1987).

Are Designers Aware of How They Design?

At the start of this chapter, design is described as being one of the higher, if not the highest, level of intellectual activity. Therefore, by implication, designers are intelligent, knowledgeable, thoughtful people, capable of reasoning about complex problems and providing solutions.

One would therefore expect that designers would have a good appreciation of design and the methods they use. Amazingly, studies undertaken of designers at work show that this is not the case and that in fact designers have a very poor understanding of their mental processes. Before we are accused of insulting all those involved with design, it must be explained that the same is true of other high-level human cognitive processes. Reference back to Chapter 2, where we discussed procedural and declarative knowledge and associated mental processes, will help to explain the above.

Evidence of this lack of understanding by designers has been collected by a number of people, for example, see the work of Wallace and Hales (1987), Baker et al. (1989), Finger and Dixon (1989a) and Moore (1991).

Protocol Analysis

Given that designers themselves are not able to explain their higher cognitive processes, what techniques can be used to find out more about design? The most successful technique that has been used is one called protocol analysis (Newell and Simon, 1972). Subjects are asked to talk aloud (sometimes described as "thinking aloud") while performing a task. The process is captured by tape recording, and where possible by filming, the actions of the subject so that as complete a record as possible is obtained. This verbal/visual protocol is then analysed in an attempt to determine the thought process used to solve the problem.

Difficulties with the technique immediately spring to mind. For example, if the designer has a very poor knowledge of his/her higher cognitive processes then he/she will be unable to articulate them. Certainly there are periods during the collection of verbal protocols when the subject sits and thinks, and although one can encourage him/her to verbalise, there will inevitably be some thoughts which cannot be captured.

The need for the subjects of the protocol analysis to verbalise tends to slow them down, but research has shown that this does not appear to alter the order or the content of their thought process (Ericsson and Simon, 1980). The fact that people are aware of being observed and that their utterances are being recorded has the potential to encourage people to alter their thought processes to ensure that they seem logical and sensible, although the available evidence shows that they are not too successful at this.

Also, design is often undertaken as a team activity. Unfortunately the use of verbal protocols is really best suited to the collection of data from one person working alone.

From the above discussion it can be seen that there are problems with the use of protocol analysis, but the general consensus of opinion is that it is the best available tool for gaining an insight into the thought processes of people undertaking tasks such as design (Ullman et al., 1987). A diagrammatic evaluation of its effectiveness is given in Fig. 4.2. The parts of the design process which Fig. 4.2 shows to be missed by protocol analysis have to be inferred. An assessment of the validity of these inferences can be made by showing them to the subject used for the protocol analysis, but it is generally accepted that retrospectively thinking about a problem-solving process elicits what people perceived to have happened rather than what actually took place, and that there can be quite substantial differences between the two (Ericsson and Simon, 1980).

So How Do Designers Design?

There have been several studies based on protocol analysis of designers at work, notably those of Akin (1978, 1986) and Ullman et al. (1987). What follows below is based on these two studies plus the work of Moore (1991). The work of Akin covered architecture, that of Ullman et al. covered mechanical engineering and that of Moore civil engineering. All of the above studies involved observations of the conceptual stage of the design.

Akin asked an architect to design a small house for a blind person. The architect spent 4 hours on the problem, developing a conceptual solution which was described in a set of 1:8 scale plans. The session was recorded on video.

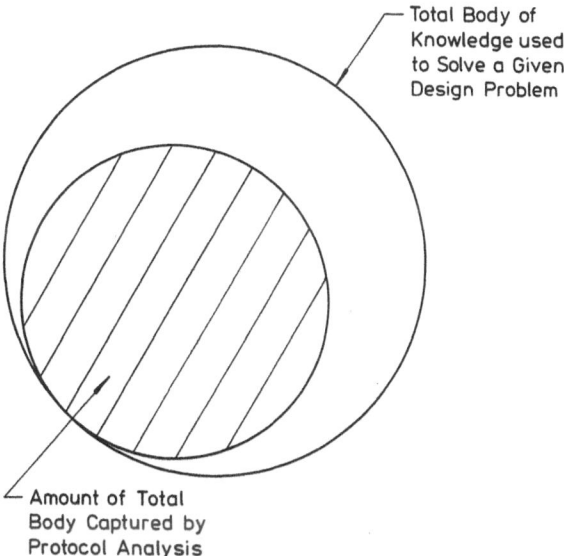

Total Body of
Knowledge used
to Solve a Given
Design Problem

Amount of Total
Body Captured by
Protocol Analysis

Fig. 4.2 Effectiveness of the design process

The architect developed three possible solutions which were based on different points of entry into the building. Interestingly, the architect recommended the solution which was based on the proposal which he considered first, and he is quoted as saying that he often finds that first ideas are the best. We will return to this point later.

In the analysis of the protocols, Akin shows how the designer uses specific information (in this case about the needs of blind people) and general information about house design and materials. This idea of general knowledge can be taken further to include "common sense" and other forms of everyday knowledge possessed by designers. Both of these types of general knowledge are very difficult to incorporate in design KBS because they are so extensive. Humans acquire knowledge over many years, in fact the process continues throughout our lives. Also Akin clearly shows the importance of heuristics in design with, for example, several statements regarding costs which were made without any apparent supporting decisions.

Ullman *et al.* (1987) tested two design problems on two groups with different levels of expertise, these being graduate students and experienced designers. Each design problem took about 10 hours to complete. This is too long for the designer to complete without a break. In practice it was found that between 2 and 3 hours was as much as could be coped with in one "session", and so the development of the entire design had to take place over several days. The participants were asked not to think about the problem between sessions and checks were made to see if this request was being obeyed, although the checks were not foolproof. The work was tape-recorded and filmed on video.

One useful feature of the reporting of this work is that a time scale is given so that one can assess how quickly given features of the design were arrived at. The main finding is that designers tend to pursue a single conceptual design, rather than evaluating options and developing these until it becomes clear which is the superior design. Ullman *et al.* found that the designers arrived at an initial conceptual idea with 45 minutes of starting, and that when deficiencies appeared as the concept was further developed, rather than consider alternatives which might have been more suitable, the designers "patched" their original solution.

This single solution strategy has been noted by others, for example, Adelson (1988), Burgoyne and Jayasinghe (1991) and Moore (1991). As shown above, the architectural designer studied by Akin did develop three alternative solutions although finally the first solution was chosen. Whether the multi-solution design strategy is common among architects and the choice of the first solution is a manifestation of the type of design behaviour observed by Ullman *et al.* cannot be inferred from such a small sample. The teaching of architects involves a lot more work on concepts than is usual in engineering, and one wonders whether this is what caused the architect to consider options?

After reaching the initial concept, the designers studied by Ullman *et al.* progressively refined their ideas, tending to work on components in turn. Only when the detailed stage of design was reached did effort tend to focus on a single item for any length of time. However, there was a lot of variation in behaviour between designers.

One consistent feature was that as the design progressed, the designers tended to abandon any logical process and pursue an opportunistic strategy with decisions being made without any apparent firm supporting case. This was found to be so even when the designer started off by using systematic generate and test procedures. Furthermore it was also found that designers (even the experienced engineers) made mistakes which they often failed to detect, and then based subsequent design decisions on these mistakes. Even when the mistakes were detected or pointed out the designers failed to check back through their decisions to assess the impact of the error, and instead carried on with flawed designs.

Some further supporting evidence for the above can be found in the work of Lansdown (1989) who states that designers, particularly during conceptual design, tend to base their design decisions on their intuition and experience rather than first trying to acquire useful task-specific information. The acquisition of the latter is at best postponed to the more detailed stages of design. As Lansdown rightly points out, decisions taken at the conceptual stage tend to have far reaching effects on the final form of the product. Therefore if these decisions are made on the basis of inadequate information then there is a strong chance that the utility of the product will be less than it should be. In defence of designers it should be pointed out that often their ability to pursue options is restricted by lack of time (Marples, 1961)

To practising designers, much of the above will be rather upsetting. No doubt most designers feel that they are doing a first-class job, and indeed they are, but all the available information on how design is actually undertaken indicates that the processes used are far less logical and consistent than one would intuitively presume. This it would appear is not due to any personal inadequacies but to the limitations of the human brain.

Memory, Notes and Sketches

As discussed in Chapter 2, humans can hold about 7 ($\pm$ 2) items in their short-term memory. Any other data either have to be stored in long-term memory or some form of external memory (i.e. notes or sketches) has to be used.

Design is a very complex process and all studies of the design process have noted that as the design progresses, designers make extensive notes and use sketches as memory aids.

Designers also store ideas and information in long-term memory. However, Ullman *et al.* (1987) noted that designers tend to forget things that have been committed to memory even to the extent that they repeat previous bits of the design process. This type of problem and the transition to opportunistic behaviour which is discussed above are speculatively attributed to the occurrence of cognitive overload (i.e. the problem is too complex for the brain to deal with in its complete form). The designer can only process items which can be stored in short-term memory, and consequently the design process tends to focus on what can be remembered (no more than 9 items) and what is in the immediate field of vision (Ullman *et al.*). Thus, sketches and notes are used by designers as a form of external memory.

Notes and sketches have other uses in design. Obviously sketches help the designer to visualise the problem; in the case of machinery they are used to help the designer mentally simulate movements and in all design (except software) understanding the geometric relationships between components can only really be achieved by drawings.

Most disciplines have formal graphic languages which are used to communicate ideas. Examples of these languages are given in Fig. 4.3. Also most disciplines have more than one language. Usually there is a set of symbols which are used for the earlier stages of design, that is concept and analysis. These symbols are usually highly stylised and of the form shown in Fig. 4.3. We will call this the conceptual graphics. In addition to this, there is then also a more formal graphical language which is used for the detailed drawings that are used to communicate with the manufacturer of the artefact. This formal language is a reasonably accurate two-dimensional representation, but certain features have to be included in semi-symbolic form.

Conceptual graphics are for many disciplines recognised internationally, thus helping to break the language barrier which occurs with spoken and written communication. Their use by designers is not just as an aid in the design process itself. When conversing with others designers, they habitually use sketches to convey form and function in a manner which it would be difficult and cumbersome to achieve with words. Thus graphical symbols are an essential supplement to other means of communication.

Sketches also act as a form of checking on the progress of a design.

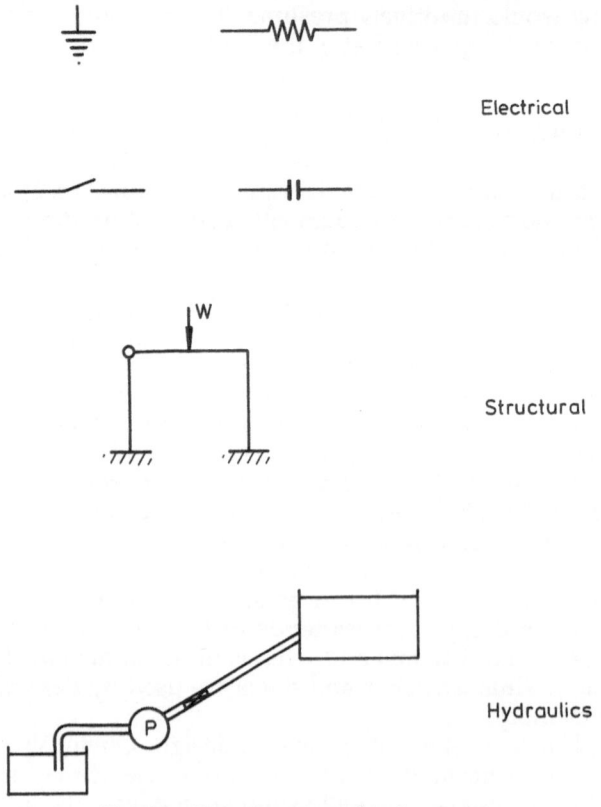

Electrical

Structural

Hydraulics

Fig. 4.3   Examples of graphics languages

Making rough drawings of what has been designed helps to reveal if any components have been omitted or are incomplete. Also, where the design incorporates moving components, sketching can help with the visualisation of the movement.

Chunking of Knowledge and Recall

The ability of the human short-term memory to store 7 ($\pm$ 2) items has already been discussed several times, but what does an "item" consist of? For example, if each item can be no larger than a single character then this is quite a severe restriction, but if each item consists of a large amount of data (e.g. an entire drawing of a component) then the short-term memory becomes a much more powerful and useful feature.

We will start off by looking at an area which has little to do with design but which nevertheless gives an insight into how the short-term memory works.

Newell and Simon (1972) discuss a series of experimental results gained from studying chess players at all levels of ability from grandmaster downwards. It was found that an expert chess player could, after looking at a chess board for 5 seconds, subsequently reproduce from memory, almost without error, the positions of the pieces. Weak players conversely could only reproduce the position of about half a dozen pieces correctly. The above findings held so long as the placing of the pieces was based on actual games of chess, that is so long as there was a logical structure to the locations. However, if the location of the chess pieces was randomised then the performance standard of the experts fell to that of the weak players, that is they were able to reproduce the position of about half a dozen pieces.

Now given the above capacity of the short-term memory and also the fact that probably not more than one item of data can be transferred from short-term to long-term memory in 5 seconds, what can we infer from the above?

It is thought that when the chess pieces are in logical places which are dictated by the rules and accepted procedures of chess then experienced players, rather than remembering the position of each piece as a single item in memory, are able to "chunk" several pieces together into a "constellation" and remember this as a single item. The logical relationship between the items helps with the encoding and decoding of the chunk. However, weak players do not have this logical relationship mapping ability, and so are unable to group the positions of pieces into chunks; hence they store the locations of single pieces, resulting in the capacity of the short-term memory restricting this to about half a dozen correct places.

The above is getting us a long way away from design, but, as is shown below, the explanation is of relevance. Akin (1978, 1986) studied the performance of architects and Waldron et al. (1987) studied mechanical engineers. We will first examine the work of Akin.

Akin asked his subjects initially to undertake tasks which were intended to implant the details of some architectural drawings (floor plans of buildings) in the long-term memory of the subjects. Three types of task were used, these being tracing, copying and interpreting. The first two tasks need no explanation, but the third involved the subject studying an unlabelled floor plan and determining the type of building and the functions of the individual spaces.

Once the subjects had completed the above tasks they were then asked to recall the drawings from memory. Not surprisingly, in the recall tests, those who had traced the drawings did worst in terms of accuracy because their task placed few demands on the subjects' memories and those who had been asked to interpret drawings were the most accurate. The line drawing tests were recorded on video and the latter were then used to study the latencies (i.e. pauses) between the drawing of each line. Akin reasoned that if the moment a line was completed the next line was begun then the same chunk was being dealt with, but if there was a pause then the subject was recalling another chunk from memory. Using this method of analysis it was found that the chunks were fairly small: corners of rooms, stairs, walls with windows and other openings. There was no indication that objects of the size of whole rooms, for example, could be stored as chunks. Unfortunately Akin did not test subjects of different levels of experience and so we have no idea from his work of how what is contained within a chunk changes as the designer becomes more familiar with the domain.

Next Akin took the logical step of trying to find out how the memory manages to store the relationships between these chunks, because not only are the chunks recalled but they are placed in a specific spatial order. Tests showed that six types of information were probably used to structure these organisational relationships in memory. The categories are: adjacency, size, orientation, function, circulation and structure. Obviously some of these are only relevant to architecture but it would seem reasonable to presume, provided the above are correct, that equivalents exist in other disciplines.

Waldron *et al.* (1987), as stated above, used mechanical engineering as the domain for their work. They used six drawings which they presented to three groups of subjects, these being:

- Expert designers.
- Semi-experts, largely graduate students.
- Students with a very basic grounding in design.

Each subject was asked to look at a drawing and then reproduce as much of it as possible from memory. The subject was allowed to look at the drawing whenever necessary, but the number of references was recorded as was the time between each reference.

Overall, as one would expect, the experts performed better than the others, although significant variations in performance within groups are reported. The average performance of the groups shows that the experts made fewer mistakes and fewer references to the drawing, although the durations of their studies of the drawing were longer. This tends to confirm what one would expect from the study of chess players, that the value of experience is that one is able to handle information in larger chunks in short-term memory.

Summary of "Real Design"

So to summarise briefly; the above description of "real design" is based on a very small amount of data. There are one or two other examples which have not been mentioned above, although the findings of these extra studies broadly agree with those covered herein, but because the amount of data is small any conclusions which are reached can at best only be tentative.

Given the above it can be concluded that:

- Designers tend to suffer from cognitive overload because of the complexity of the design task. This can lead to opportunistic behaviour.
- Designers tend to work with their initial conceptual idea and "patch" this to obtain a workable solution rather than evaluating a range of options.
- Designers are not aware of their higher-level cognitive processes.
- The capacity of the human short-term memory has a substantial influence on our ability to undertake design.

- Increasing the designer's expertise enables his/her brain to handle information in larger "chunks".
- Sketches serve a variety of purposes, helping to visualise the problem, assess solutions, providing an external memory and providing a means of communication between colleagues.

Systematic Design

Introduction

In an attempt to give some structure and organisation to the design process, many people have described what they believe to be the ideal way to undertake design. Most of these examples are from German-speaking countries and predominantly from Germany itself. With the increasing complexity of products and also recognising that when designs are undertaken by teams, some form of common strategy is necessary, it is argued by the advocates of this type of design process that following a well defined path leads to better designs. The underlying reasoning behind these structured design processes is called design theory, and the processes themselves are called systematic design.

Many examples of design theories are available in the literature (for example, see Pahl and Beitz, 1988 (original German version 1977) and Hubka, 1982 (original German version 1980)). Each contribution advocates its own unique design theory (Lansdown, 1989) but the overall aim of each theory is the same, to put forward a logical strategy for designing a product which is aimed at ensuring that a well thought out, reasonably optimal and economic design results.

Usually when describing a topic it is best to give examples. To accomplish this for design theories it is sensible to concentrate on one particular design theory to avoid the confusion which might arise if various theories were to be considered. In this section, therefore, we concentrate on the work of Pahl and Beitz (1988) which is arguably the best of its type. The description does not describe in detail the methods of Pahl and Beitz because it would be foolish to repeat what already exists. Instead the following concentrates on what constitutes design theory and what the implications are of following a systematic design procedure.

Origins of/Motivation for Systematic Design

As discussed above, design evolved from craft-based roots and thereafter gradually developed as products became increasingly complex during the early 19th century. According to Pahl and Beitz (1988), the initial development of modern design science came some 100 years later in the 1920s and the subject has progressively developed since that time.

The procedures recommended by the various proponents of systematic design are not based on any explicit fundamental understanding of the cognitive processes involved in design but on the intuitive knowledge that one gains with practice. It has already been stated, designers are not fully aware of their higher cognitive processes but, as is shown below, systematic design does attempt to address some of the problems caused by human fallibility.

Systematic Design: an Example

In the view of Pahl and Beitz (1988), systematic design has four components, these being:
- Clarification of the task.
- Conceptual design.
- Embodiment design.
- Detail design.

Clarification of the Task

The clarification of the task is a vital procedure and yet it is one which sometimes does not receive enough attention. The work of Ullman *et al.* (1987) shows that the designers commenced the design task without fully understanding the brief and, for example, anecdotal evidence from the software industry (Montgomery, 1992) indicates that much of product development time is wasted because the client's requirements and wishes are not fully comprehended or analysed.

Just what is to be clarified depends on the nature of the relationship between the designer and whoever initiates the demand for the design, but some general points arise:

1. All constraints/requirements should be fully investigated and understood. The number of constraints/requirements should be kept to a minimum to allow the designer maximum freedom. Attempts should be made to identify and remove "false" constraints which will at a later stage prove to be unnecessary as these can restrict the design process and result in an inferior product. A clear distinction should be made between those constraints/requirements which are mandatory and those which are wishes.
2. The designer and whoever commissions the design should both be absolutely clear about what is required of the final product and how it is intended to achieve this. For products which are to be put on sale, both sides should have a good appreciation of the needs of potential purchasers. For other types of products, the needs of users must be understood.
3. Information should be gathered about instances of similar products. The failings and successes of these products should be analysed, and steps taken to incorporate the good points and avoid the pitfalls.

Conceptual Design

There are some small differences between what Pahl and Beitz (1988) call conceptual design and what has been similarly referred to in this book.

Part of what Pahl and Beitz call embodiment design would in our definition be conceptual design, but these classifications are not important except as shorthand for sections of the design process. In this section the
Pahl and Beitz definition of conceptual design will be used. The methods
that they use are complex and will only be described here in broad outline.

As one would expect with systematic design, the whole process is formally defined. This is intended to maximise the chances of obtaining the
optimum solution and at least to ensure that a reasonable span of options
is considered.

The first decision that Pahl and Beitz say should be made is whether or
not a conceptual design is needed. If the product is such that satisfactory
known solutions exist then the process can progress directly to the
embodiment and detailed design phases. Oddly enough, given that much
of their work involves using formal methods for decision making, they do
not present a method for the above decision. It is conceivable, for example, that new technological advances have rendered the known solution
less attractive and that unless the design is changed then the product will
gradually become obsolete. While techniques are presented for dealing
with the latter, the evidence of obsolescence usually occurs some time
after the technological advance.

The first stage of the conceptual design process proper occurs with
what is called abstraction, which initially involves:

1. Eliminating personal preference.
2. Omitting non-essential requirements and constraints.
3. Transforming quantitative data into qualitative data.
4. Formulating the problem in solution neutral terms.

Having completed the above and identified the crux of the design task,
the next process is what is called a systematic broadening to discover if
there are any extensions to or even changes in the original task which
might lead to a more promising solution. At this stage it is important that
all solution paths are left open and as many solutions as possible are
generated and considered. Thus the designer must continually question
the imposed constraints to see if they can be removed. Examples are
given, such as rather than designing a garage, door why not look for a
means of securing a garage in a way that protects a car from thieves and
the weather.

An essential part of the Pahl and Beitz (1988) conceptual design procedure is the function structures. This is a way of expressing the design in
an abstract form so that concepts can easily be reasoned about because
the problem is formulated in a simple and clear manner. The method
starts by identifying the overall function and then splitting the problem
down into subfunctions which are linked by factors such as flows of energy, materials or signals. The level to which the problem is broken down
depends on the circumstances of the particular design. The objective of
the process is to facilitate the subsequent search for solutions and allow
the clear identification of existing components or design knowledge
which form a suitable starting point for the design. From this it is then
possible to say what new items need to be designed.

Once formulated, the function structures are then analysed using logical and physical relationships between the functions. The process helps to identify further the essential structure of the problem.

The next stage is to search for solutions which fulfil the subfunctions. This process is intended to lead to a number of solution variants. Several strategies to aid with this search are suggested. These come under three major headings:

1. Conventional aids: literature search; analysis of what occurs in nature; analysis of existing technical systems; analogies and measurements of existing systems and model tests.
2. Methods with an intuitive bias. These are frequently used by designers with the answer coming in a flash of inspiration after some reflection, and although there is no evidence to support this, it is generally believed that most really novel ideas are the result of intuition. Methods with an intuitive bias have the disadvantage that the inspiration may not arrive at the right time. They are also prone to personal bias (for further information on inventiveness see Dixon, 1966). The methods in this section include: discussion with colleagues; brainstorming; method 635 (a development of brainstorming); the Delphi method (based on experts being asked for written opinions); synectics (a more structured sort of brainstorming).
3. Methods with a discursive bias. These provide solutions using a more deliberate step by step approach and include: systematic study of physical processes (this is only applicable to situations which can be represented algorithmically (i.e. by equations) so that the system can be studied by altering terms within the equations); systematic search with the help of classification schemes, usually using two-dimensional data structures; use of design knowledge such as manufacturers' catalogues, design codes and text books.

Having found solutions to the subfunctions, these then have to be combined to fulfil the overall function. The main problem with this is to provide solutions which ensure physical and geometrical compatibility. This should lead to a number of solutions, especially as the above process is designed to generate as many solutions as possible and to keep the range of solutions as broad as possible. This is both a strength and a weakness of the systematic approach and Pahl and Beitz recommend that at this stage the number of solutions should be reduced to manageable proportions, otherwise the workload becomes excessive.

The number of solutions is reduced by first eliminating all the obviously unsuitable proposals. If the number remaining is still large then those which are patently better than the rest should be retained and all others discarded. Should the reduction process prove to be difficult then a systematic selection method based on a decision matrix is suggested. This stage of the process should result in a small number of feasible options which can then be evaluated in further detail. The function structure is inadequate for this more detailed assessment and so the proposals need to be consolidated and transformed into a more concrete representation (called a concept variant). This work often requires further data collection to fill in the gaps which appear.

Pahl and Beitz (1988) insist that the evaluation of the concept variants should not be made on purely financial criteria because technical factors can also play an important part in determining which design will lead to the best product. A method called "use value analysis" is applied to compare concept variants. Each concept variant is rated for its technical and economic merit, and then these two ratings are combined in one of several ways to give a final rating.

Embodiment and Detail Design

Pahl and Beitz (1988) treat these as two separate parts of the design process but as this book is devoted to conceptual design, the application of systematic methods to embodiment design and detail design will be covered in outline form only.

Embodiment design starts with the chosen concept and involves working up this idea by the use of general layout drawings to clarify the relationships between various parts of the design. The process is iterative and involves many checks to ensure compliance with design standards, safety requirements, manufacturing/construction constraints and also to ensure that the final result is a reasonably optimal solution. The techniques used involve a movement from qualitative to quantitative descriptions of the proposed product using alternately analysis and synthesis. The three key words expounded by Pahl and Beitz are clarity, simplicity and safety: good concepts for any designer to follow.

Detail design then follows on from the embodiment, which results in descriptions of the product that allow it to be manufactured/constructed.

Systematic Design: a Summary and a Comparison

Systematic design provides a structured procedure for the design process. Wherever possible a formal, well-defined path is followed. Many of the procedures are somewhat lacking in rigour in that they are not guaranteed to produce the ideal solution, but then design is such a complex subject that it would be impossible to produce a perfect design method. Despite the lack of rigour, systematic design does provide a framework in which the implication of decisions can be evaluated on a consistent and reasonably transparent basis. Similar systems are used in management where again the problems are too ill defined to permit absolute definition, and yet some form of consistent decision aid can be of use.

It is interesting to compare the previous sections on unstructured design and systematic design, and see the difference between the protocol analyses of actual designs and the methods suggested for systematic design. Unfortunately, protocol analyses of people using systematic design are not available, the only study of the application of systematic design procedures described in the literature being that of Wallace and Hales (1987). There is a hint in the work of Ullman et al. (1987) that designers who use systematic design procedures may still suffer the same problems of cognitive overload experienced by those who use more intuitive procedures. Further evidence on this would be of great use, although in the absence of firm data one can conclude that systematic design does not simplify the design process and so it is probable that cognitive overload will still occur. Indeed, if anything, systematic design makes the design process more complicated and so one could infer that its use will increase the risk of cognitive overload.

The contrast between what is apparently essentially an intuitive approach and the highly structured approach of systematic design is very striking. One could argue that systematic design is so structured that it stifles initiative or creativity, but when looked at in detail it appears to encourage the generation and consideration of new ideas. One could also argue that systematic design does not fit in with existing design practices and therefore it will be difficult to assimilate. However, when one looks at many other areas of activity, for example, air traffic control and quality assurance, we accept that there should be systematic procedures and so possibly some form of structure should be imposed on the design process.

Possibly the most telling argument against systematic design is that it inevitably involves more time, and hence money, because the consideration of many options is a lengthy process. This is undoubtedly true. Many design organisations obtain their work on the basis of fee competition, and systematic design is uneconomic for work gained in this manner. However, if those who commission designs become a little more far sighted then there is a good future for systematic design. Usually in comparison to the overall costs of a scheme, the design costs are a relatively small part and yet they have a fundamental bearing on the overall cost, durability, serviceability and utility of the product. To put it simply, a poor design results in a poor product which ultimately will cost either the manufacturer and/or the owner a lot more than it would have done to have paid for a more thorough design.

Design and KBS

Positive Arguments

In this section we will use the evidence presented above to examine the case for using KBS for conceptual design. As much is made in this book of bias and one aspect of this is confirmation bias, in the succeeding section the arguments against the use of KBS will be examined. Whether the cases for and against are argued with the same vigour will have to be determined by you, the reader!

Looking first at the psychological aspects, it has been shown that the design process is so complex that cognitive overload occurs, or to put it more simply, our brains are unable to cope with the mass of data which is generated and hence mistakes occur. Computers are very good at storing large amounts of data, so a properly formulated design KBS should be able to reduce the problems caused by cognitive overload. Simply undertaking the design with the assistance of a well formulated design system should ensure that all the stages of the design are recorded, and this should prevent work being repeated, help with clashes of information and identify inconsistent design decisions.

Another psychological aspect which has been discussed in this chapter and Chapter 2 is the problem of bias. People prefer to keep to what they have experience of, and even then they tend to be biased more towards their more recent experiences. Furthermore, Ullman *et al.* (1987) in particular showed that there is a tendency to use the first concept even when subsequent developments indicate that there might be more suitable solutions. It is possible to overcome many of the problems of bias using design KBS, and likewise the consideration of options is also something which would be greatly enhanced by using a KBS. The main danger is that the constructor(s) of the knowledge base may be guilty of bias when they construct the system. Moore and Miles (1991a) have shown that it is possible to mitigate such problems by the use of multiple sources of knowledge.

There is considerable pressure on design practices to cut costs. Often work is won by means of fee competition. This inevitably means that the time which can be allocated to a design is going to be limited and the pressures on the designers to achieve a suitable design within the given constraints can be quite severe. At the same time, technological advances mean that designers are sometimes required to work with materials/components with which they are not particularly familiar, thus increasing the difficulty of the design task. To some extent, design productivity can be improved by having people specialise in particular areas so that they become very knowledgeable and hence able to perform at high speed. This, however, leads to monotony and boredom. Design KBS offer a way of increasing productivity, especially that of less experienced engineers. Possibly, therefore, a combination of KBS and designer may manage to restore some interest and variety to the design task.

There are other pressures on designers, particularly those regarding quality, especially quality assurance, and the need to ensure consistency. As has been discussed above, the human brain is a very impressive but fallible reasoning device. Computer-based design systems are far from being perfect, indeed as the common adage states they are only as good as the worst programmer, but one of their advantages is that they are consistent and they never forget. So if a fault is found in a computer system, it can be rectified and that mistake should then never occur again. Thus computer systems can be improved, just as humans can learn, but with the important difference that humans may also forget.

Some of the consequences of design mistakes (all made by humans) are briefly reviewed by Lansdown (1989) who quotes figures that indicate that between 33% and 50% of defects in UK houses are due to inadequate design. It is stated that most of these failures occur not because the designers are working in an area where the current state of knowledge is deficient but generally because the designers have failed to locate and use the relevant knowledge.

In some industries, particularly construction, the links between the designer and the constructor are very weak and so there is little feedback to the designer regarding the problems that occurred when trying to execute the design plan. Difficulties which arise on the construction site are resolved by the people on the spot, and often the designer is blissfully unaware of the problems and therefore continues to make similar mistakes in subsequent designs. Obviously some form of feedback to the designer is needed from those who have to make the artefact. This could be achieved by a number of means including KBS.

One of the major stores of knowledge in any design practice is the record of past designs. Currently a few companies have design databases but most just rely on the memory of the senior engineers to find designs that are similar to the work in hand and which might form the basis of the proposed product. Database technology and KBS technology are closely linked and in the future it is probable that these links will grow stronger. A database of past designs would form a useful adjunct to a design KBS, providing a vast reservoir of knowledge. The major difficulties in using such data would be in knowledge capture, particularly if the data are stored on paper, and in the diversity of forms of representation ranging from text and drawings to computer programs. The probable way forward is by the use of object oriented database technology which although as yet not fully established as a practical tool has great potential. An ultimately more promising approach than just a database of designs is case-based reasoning where rather than just storing design information, the system uses the data to reason about designs. The technology of such systems is being developed but is not yet suitable for practical applications (Mostow and Barley, 1987; Sycara and Navinchandra, 1989).

As already discussed, in the English-speaking world design teaching concentrates on analysis with most other aspects being only briefly mentioned. To change embedded attitudes takes a long time. In the meantime it is necessary to compete with those economies where design is better taught. Design KBS offer a way of filling the void and introducing a sensible structure to the conceptual design process rather than just relying on the designers' experience and common sense.

At the moment there is a gradual transition from traditional processes to those which are based on digital technology. For example, in the construction industry, details of a construction site may be given using the traditional paper format or it is now possible for them to be in machine-readable form. The latter is increasingly set to become the norm and so in the future much of the incoming information for a design will be computer based. It therefore makes sense to continue the design process in this medium, possibly using some type of CAD-based format (in its narrowest sense of drafting) to sketch out ideas. Coupling KBS to this type of system would in concept give a very powerful tool in which the designer inputs his/her own thoughts to the design and the KBS acts almost as a colleague, offering alternatives or advice and help when necessary. This approach has the advantage that the data can be transferred to the manufacturer/constructor in machine-readable form, making the whole process very efficient.

To make the above viable, however, it is necessary to develop product models which will give some form of common representation of knowledge between different software systems. Work on this is in hand, although progress is slower than one would desire (Watson, 1992).

Negative Arguments

Some of the thoughts given in this section have been discussed elsewhere in the book, but it is useful to draw them all briefly together and examine them in their entirety.

A common fear is that design KBS, in which a large fraction of a domain's current design knowledge is stored, will stultify the development of new designs or products. One possible counterargument is that books have not had this effect, so why should KBS be any different? Nevertheless, one has to recognise that KBS are much more powerful than books because they lead the user directly to a solution. However, humans are innately curious and one only has to watch the performance of someone on a computer game to realise that over a period of time they get to the stage where they fully understand the game, have spotted its limitations and are looking to move on to something better. It is probable that with design KBS the users will, over a period of time, assimilate the information in the system and then start to develop their own expertise.

In addition, the current form of "expert systems" has been developed for diagnostic type problems. It is possible that for design KBS their evolution will result in something which allows a far greater degree of flexibility and input from the user, so that the designer controls the process.

Another problem with knowledge bases is that they represent the knowledge of at most two or three experts. What happens if these particular experts have a bias towards one type of solution, won't the KBS just replicate this bias? The simple answer to this is yes, but then this is the situation that persists at present with human experts. A well constructed KBS should help to eliminate bias (see, for example, Moore, 1991) and provide a well reasoned solution.

The dichotomy that occurs with regard to mistakes has already been covered. People readily accept that humans make mistakes and yet they are happy to let these fallible beings do such risky things as brain surgery. It seems that because we recognise our own fallibility, we are willing to accept it in others. Yet one of the major arguments against the introduction of KBS is that they cannot be guaranteed to be fully reliable.

One recognised problem with KBS is that they only cover a narrow domain of knowledge and when they are asked to step outside this domain their performance degrades rapidly. This is a real problem with KBS. Work is in hand to try to overcome this difficulty, but as yet little real progress has been made.

A related problem is that KBS lack common sense. Humans acquire this invaluable store of knowledge over many years. It covers many aspects of life and enables them to be very flexible in their thinking and application of solution strategies. Given current technology, there is no way that a KBS could replicate human performance in this area. The ways of avoiding this difficulty have been covered above. Design KBS should be designed to complement not mimic human performance, and should allow the designer to mobilise fully his/her own intelligence. If using a KBS holds back the designer then that system will never become a useful tool, and hence will not be used.

Some people have an in-built prejudice against computers. One sees this with teaching undergraduate students – some people just hate computers while others take to them without any problems. Work is in hand to try and improve the acceptability of the interface of design KBS (see, for example, Philbey *et al.*, 1991) but it is likely that some people will always be uncomfortable with computers. If design KBS become sufficiently useful that they form part of the everyday tools of the designer, then people who are antipathetic to computers will not be designers.

Finally there is the "Sci-Fi" fear that computers will take over all the thinking and that there will be nothing left for man to do. Such thoughts make good fiction but at present the sort of systems which are feasible are a long way from supplanting man.

The Place of Design KBS

The future of design KBS is dealt with in the final chapter but it is useful to summarise briefly what such systems have to offer:

1. The ability to evaluate rapidly options during conceptual design in much greater detail than it is currently economic to attempt.
2. A source of alternative solutions and hence a means of avoiding excessive bias.
3. A defined design method which will help to ensure that all aspects have been explored and evaluated in sufficient detail.
4. A help facility for those who either require inspiration or who are inexperienced or lacking in knowledge about a particular area.
5. A checking procedure for faults.
6. A means of assisting designers with the cognitive overload that currently occurs.

Chapter 5

Knowledge Acquisition and Interpretation

Objectives

The majority of practical KBS are built using knowledge obtained from domain experts and documents. Unless the expert(s) and the knowledge engineer are the same person (which, incidentally, is not always a good idea), information and expertise on the target domain must somehow be obtained in order to build a KBS. This process is called knowledge acquisition (KA).

Once the knowledge has been acquired, it must be analysed and interpreted so that it can included in a computer system.

These topics form the basis of this chapter, which aims to:

- Emphasise the importance of assessing your target domain.
- Discuss the task of finding suitable experts; whether to use single or multiple experts and how to assess these experts.
- Compare various approaches to knowledge engineering and describe possible interviewing techniques.
- Discuss the analysis and interpretation of the knowledge.

Introduction

Knowledge acquisition is recognised as being one of the most problematic areas of KBS development (Welbank, 1983; Greenwell, 1987; Kidd, 1987). This is not surprising as it is an area which encompasses a number of skills including the use of AI techniques, philosophy and psychology (Kidd, 1987).

There is often some confusion between the terms knowledge acquisition and knowledge elicitation, and the two are frequently taken to be interchangeable. In this book, they are distinguished. Knowledge acquisition (KA) is taken to be the complete process of obtaining information on the target domain: by using text books, Design Standards and Codes of Practice, other available literature and existing computer systems in addition to the experts themselves. Knowledge elicitation (KE) is the subsection of KA which covers the acquisition of knowledge by interaction with experts, thus aiming to capture human expertise directly. This is generally the most problematic section of KA and therefore dominates this chapter. This is not to say that the knowledge obtained from other sources is not equally valuable in KBS development. It is merely recognised that more problems are usually encountered when dealing with people.

These problems can be numerous and are generally well documented. In this chapter we highlight many of the difficulties, which may develop and suggest ways of dealing with them.

Once again, the discussion in this chapter is based on our own experience. As well as describing feasible approaches to KE, we encompass the less well documented parts of the KE process such as choosing and assessing the experts.

Assessing Your Target Domain

Before beginning the KE, you must ensure that the project's target domain is suited to KBS development. This may seem obvious, but often it is not until the domain is scrutinised that potential problems are identified (see Chapter 9 for an example of this).

The qualities which constitute a suitable domain are well discussed. They include:

1. Focusing on a narrow specialised area which does not involve a lot of common sense knowledge (Alty and Coombs, 1984).
2. Selecting a task which is neither too easy nor too difficult for human experts.
3. Selecting a task which does not take an expert more than a few hours to solve.
4. Selecting a domain in which the usefulness of encapsulated expertise is obvious.

The most important factor is that the necessary expertise is available. It can be very frustrating to start a project only to find that no actual "expertise" exists or that the existing expertise is inaccessible. If there are only a few experts in the world, how feasible is it to obtain information from them?

Don't just consider the experts either. What about the resources you, as a developer, have available? Have you the time, software, hardware and support required to develop the system?

If it is concluded that the overall domain is suitable, the chosen area needs to be looked at more carefully. Is the domain over-ambitious and is it really likely that the system will do what you claim? Decide the limits of the domain in conjunction with the experts: after all, they should have a good idea of which aims are sensible and which will be most useful. For example, when developing the bridge design system (Moore, 1991), it was decided in conjunction with our experts that to try and deal with all types of bridge was over-ambitious and that it was better to limit the system so that it only dealt with small to medium span road bridges crossing another road. It is better to produce a smaller but successful system than to risk failure by being over-ambitious. For the bridge system, small to medium span road bridges were chosen as these are the type most often encountered by young inexperienced engineers – the targeted users of the system. After all, a system can be expanded if the initial version proves to be effective, and extended co-operation from the experts is more likely if the first part of the project is successful.

Identification of Users and their Requirements

Never forget the users who have been targeted. This is a point which is accentuated repeatedly in this book, as we feel that user recognition is an important criterion in KBS development, as it is where many practical systems fail. Look carefully at the benefits which will be gained by developing the system – whether these be in terms of cost, quality or time, and who will actually use the system. The system does not have to be capable of replacing the original expert: most developers would agree that this is currently not possible. However, if the system is not going to be of real use, why develop it?

From this information, decide who the likely users will be. Once these users have been identified, talk to them and find out if they would find the system useful. If their reaction is not positive, either the aims of the system or the intended users need to be reassessed. If the users like the idea of the system, find out in more detail what they would like the system to do. Keep the users involved throughout the development and evaluation of the system. They should not affect the expertise which is included, but they should influence the aims of the system – the way it is structured and the facilities which it provides – to ensure its applicability.

Another way of ensuring that the system is developing to suit its intended users is to study the working practices of those users. This enables the applicability of the system to be assessed, so that the system can be altered to suit what the users actually do and the way in which they process their work.

Choosing the Right Experts

Finding the right expert(s) for the KE is a vital part of building a KBS. Without these invaluable people, it is unlikely that the project will develop beyond the initial stages.

Can the System Developer be the Expert?

Some systems have been developed by the experts themselves (Milne, 1990; Soh, 1990). However, this is not always a good idea. It is very difficult to analyse effectively the domain expertise yourself, making it hard to develop an objective and useful system. This is because much of the knowledge that differentiates pure knowledge from expertise is procedural in nature, and thus intrinsically difficult to identify and articulate. Identifying how to do something is much harder than identifying what that something is and, particularly with design knowledge, knowing how the design is carried out is the important part of the KE process. It is hard enough to identify how someone else is carrying out a task without having to analyse your own expertise. For example, driving a car: it is much easier to teach someone to drive by identifying their mistakes than it is initially to explain to them the driving processes which you actually use yourself.

The importance of bias has been discussed elsewhere (see Chapters 2, 4 and 8). Eliciting an expert's biases is extremely difficult, but identifying your own biases is virtually impossible. We all prefer to believe that the judgements we make are objective: unfortunately, this is frequently not the case.

For a more detailed explanation of the reasons why procedural knowledge, especially your own, is hard to identify (see Chapter 2).

What Makes an Expert "Right"?

The "right" experts have to possess a number of qualities: they have to be articulate, co-operative, sympathetic, enthusiastic, flexible and a genuine expert. These people are inevitably hard to find. You may well have to compromise on some of these qualities in order to find an expert, but in our experience time is better spent searching for an alternative expert than wasted on unproductive KE.

The problem of finding suitable experts can be split into two stages: firstly, finding a company or companies who are willing to allow their employees or partners to co-operate, and secondly identifying the individual experts.

Finding Co-operative Companies

This is not as straightforward as it sounds: we have encountered companies whose experts were interested in the project but whose directors were unco-operative.

Companies are naturally concerned about revealing the experience upon which their livelihood is based, especially in a competitive field such as engineering. Understandably, the concept of giving away their knowledge to competitors free of charge is far from appealing! This is a particular problem when dealing with design domains, as the design expertise they possess is the only thing which makes them a better designer than their opposing companies. They are understandably reluctant to reveal such important expertise.

A second deterrent is the amount of time which the companies believe they would have to commit to the KE. In fact, the time which an individual expert is actually required to devote to the KE is usually very small. However, some companies are still of the opinion that this time could be put to more profitable use, feeling that they would be providing something for nothing: their time and expertise for no financial returns. Generally, the only reward which can be offered is the completed system, and often this is not a sufficient incentive.

If the project is "in house" then obviously this argument is not relevant, as the company must have sufficient confidence in both the idea and the success of the completed system to sanction the project in the first place. Persuading them to commit a large amount of time to the project may still be problematic, but with clever management, it should be possible to acquire the necessary amount of time from the experts involved.

Finding the Individual Experts

Finding appropriate experts is a very common problem in KE (Fox *et al.*, 1983; Grover, 1983; Smith and Baker, 1983). Committing time and effort to finding a suitable expert can make the difference between a project's success or failure (Welbank, 1983).

Unfortunately, not all experts are suitable. This unsuitability is generally not a reflection on their expertise. Some of the main causes are discussed below.

The Inarticulate Expert

Some experts find it extremely difficult to formalise the problem-solving techniques they use (Ericsson and Simon, 1980) and often experts solve problems in a way which is difficult to verbalise. This can sometimes be overcome by encouragement on the part of the knowledge engineer, but unfortunately if the expert is truly inarticulate in terms of explaining his/her problem-solving techniques this can often prove to be unproductive. If the expert finds explaining his or her reasoning processes very difficult, and struggles to do so, this will only succeed in making the KE more tedious than necessary for both the knowledge engineer and the expert.

The Sceptical Expert

One of the most unproductive interviews which we have experienced was basically because the expert did not see the point in carrying out the interview, as he did not believe the system would succeed. The expert was very sceptical about the feasibility of the project, and consequently proved to be totally unco-operative.

Sometimes, experts refuse to see the relevance or future potential of a KBS, or refuse to believe that the system will ever be constructed. This attitude can be directly stated or it can be inferred by the expert answering questions in a derisive manner, in an attempt to show that his expertise cannot be simulated. Some experts do not believe that any section of their expertise could ever be captured and emulated by a computer. This is also generally reflected in their attitude to the KE.

Careful explanation of the project's intentions can sometimes help to persuade the expert that the project is viable and worthwhile, but unfortunately occasionally experts will be encountered who refuse to be convinced, even after sustained encouragement. In these cases it is better to look for an alternative expert, as it is unlikely that the information acquired from the expert will be sufficient to build a viable KBS.

The Expert Who Fears Replacement or Redundancy

One of the reasons experts are unco-operative is because they see the KBS as a threat to their job. These experts may feel that by protecting their knowledge, their position will remain secure. This can be overcome by stressing that the KBS is not meant to replace expertise, merely support it. In many cases, it may help to state that in most organisations, it is not those at the top who are threatened by computerisation, but those further down the chain.

The Non-Committal Expert

Sometimes experts are encountered who are unwilling to commit themselves to decisions or statements, preferring to make generalised remarks. Such people can make KE a very unrewarding and frustrating experience.

There can be a number of reasons for this non-committal behaviour, including scepticism, lack of confidence or fear of replacement, all of which have already been discussed. However, there is another possible cause: when the expert sees the developer as a potential client. In this situation, the expert refuses to show a preference in case it jeopardises potential business.

An example of this was encountered when we were developing the bridge design system. A construction industry contractor was interviewed as part of the KE process, as we felt that this would help to identify the on-site difficulties which are caused by poor designs. It was found that the contractor would not state a preference for any design or form of construction, only stating that anything was design possible, if it was required. We got the impression that the contractor was treating us like a client and he did not want to admit to any preferences in case it risked losing trade. Despite attempts to break down these barriers, the expert was unwilling to say that in practice a certain type of design would be preferred.

The Out-of-date Expert

In a field such as design, expertise has to be maintained. Generally, designers need current practical experience if their expertise is to be relevant and useful, otherwise there is a risk of their knowledge becoming out of date. This is a problem which we have encountered: one of the experts which we have used had been an established and well respected designer, but in recent years had moved into different areas of work, practically retiring as a designer and taking on a more administrative role, and consequently his design expertise had diminished. We could not use this type of expert for the KE as it would have been impractical, as his expertise would need to be supplemented by experts who were more in touch with current developments.

The Inaccessible Expert

Unfortunately, the problem of inaccessibility will be somewhat apparent regardless of the expert chosen, as often expertise is inextricably linked to demand.

Inaccessibility is a particularly difficult problem to overcome with designers because, as mentioned earlier, to remain expert in design-sustained practical experience is required. Also, expert designers invariably hold senior positions within their company. Therefore, expert designers still have many commitments to their design work as well as to their position in the company, meaning that very little of their time is free. This results in experts being frequently unavailable owing to trips abroad or general work pressures, often causing arranged meetings to be cancelled at the last minute. This problem is true of most expert designers, so unfortunately, using an alternative expert is of little benefit.

Using more than one expert can help to overcome the problem of inaccessibility, as is discussed later in this chapter. The only other solution is to be patient and, moreover, be flexible. Be prepared to see experts whenever it suits them, either in work hours or out, and do try to keep the interviews short so they will be more willing to see you the next time. Also try to meet in a place where there will be no interruptions, so that maximum use can be made of the time which is available.

The Expert Who Doubts His/Her Own Expertise

This is another problem which we have encountered with an expert which we have interviewed: meeting an expert who doubts their own ability can be very difficult! Experts who are insecure in their position tend to hide their expertise by being non-committal or by being generally evasive when questioned on their expertise. They also regularly refer to standard "text book" answers, in an attempt to prove that they are knowledgeable in their subject. Some may even become very defensive, refusing to co-operate at all during the interview. Unfortunately, experts who are not in fact real experts tend to exhibit the same characteristics as experts who doubt their own expertise. There is therefore a great difficulty in distinguishing the expert who is insecure in their expertise from the experts who are trying to cover up their lack of expertise! The only real solution is to trust your own judgement.

Once you feel that the expert you are dealing with is a true expert who merely lacks confidence, then continued encouragement and confidence boosting can overcome the problem. Once the expert realises that you are not set on revealing his or her lack of expertise, they generally become more relaxed with the entire KE process and they become more co-operative. Some even get to enjoy exploring their own expertise as it helps them realise their strengths and hence gain confidence in their own ability.

The "Fake" Expert

These are a particularly difficult "experts" to identify. As mentioned in the previous section, they can exhibit a number of characteristics which may be due to reasons other than their lack of expertise. However, sur-prisingly enough, in time, "fake" experts usually reveal themselves. Key signs are if experts are more managers than "hands on" people. This tends to be so, especially in design: if a designer is a good designer he/she will rarely relinquish contacts with the design process altogether. We have met a number of extremely good designers who have graduated to management but who still maintain their expertise. If a person seems to have cut all ties with the design process, it is usually a bad sign.

One of the best indicators of false expertise is when the expert tends to become aggressive when repeatedly asked to explain a solution. Also, if an expert is being particularly evasive, refusing to be pinned down on anything, this may be suspicious. In any case, once you have identified the expert as a fake, realising that they do not possess the expertise which their reputation claims, inevitably it is better to move on and find an expert who lives up to his/her reputation!

Single or Multiple Experts?

In the past, it has been generally accepted that to construct an effective KBS, only one expert should be used for the KE (Dixon and Simmons, 1983; Hayes-Roth et al., 1983). This single expert approach has been so well accepted that the alternative of using more than one expert has not been considered (Chung and Kumar, 1987; Fox et al., 1987; Inder et al., 1987; Trimble and Cooper, 1987; Welbank, 1987). The standard argument against using multiple experts is that it would be detrimental, leading to confusion and conflict within the knowledge base (Hayes-Roth et al., 1983; Greenwell, 1987; Alvey and Greaves, 1990; Bramer, 1990; Keen, 1990).

McGraw and Seale (1988) do discuss the use of multiple experts, but only in the context of getting people of a lower relative status to contri-bute to discussions. Therefore, effectively, they were not interviewing multiple "experts" only a number of people who were involved in the subject. Also McGraw and Seale (1988) only concentrated on the brainstorming approach: only interviewing multiple knowledge sources in groups and not attempting to deal with each expert separately. This contrasts with the approach which we adopted, where we interviewed each expert individually.

Mittal and Dym (1985) have also discussed the concept of multi-expert KE. They have suggested that using more than one expert is useful when a domain contains a large number of subject areas and a shared level of expertise exists. That is, different experts exist for different areas of the domain. Interviewing more than one expert thus ensures that an expert for each subject area is analysed.

Using more than one expert has not been given the consideration it deserves. In our work, we have used multiple experts for the entire KE process, not because of shared expertise in the domain, nor for the reasons discussed by McGraw and Seale (1987), but to improve the efficiency of the KE (Moore and Miles, 1991a). As mentioned earlier, we interviewed each expert individually and did not rely on the brainstorming technique. The apparent advantages and disadvantages of using multiple experts are discussed in the following sections. This discussion shows that the advantages can outweigh the disadvantages and that this approach for KE should be given careful consideration.

However, inevitably, the circumstances of the project must be assessed before a decision on how many experts to use is reached. For example, if only one expert is feasibly available to you, then there is very little choice; or if the system is being developed in house to supplement an individual's expertise, then a single expert is sensible.

Advantages of Using More than One Expert

Our experience in KE has shown a number of advantages of using more than one expert for the KE. The main ones are discussed below.

Reduced Inaccessibility

As has been mentioned earlier, inaccessibility is a common problem, particularly when dealing with design experts. The problems which we have encountered in the past in this area have been considerable, mainly owing to the pressures of the workload which the experts were experiencing. During one project in particular, the problem of inaccessibility became so bad that we felt that steps had to be taken to overcome it, as it was affecting the progress of the project. Therefore, we chose to use more than one expert, hoping that this would help to ease the problem.

This approach was found to be effective in reducing the expert inaccessibility. This is an immediate advantage of using more than one expert. By using more than one expert for the KE, the problem of all of these experts being simultaneously unavailable is rare. As inaccessibility is such a common problem, anything which helps to reduce it is welcomed.

Alternative Source for Explanations

When one of the experts finds it difficult to articulate the reasoning behind a decision, usually one or more of the other experts are able to provide an explanation about how this decision is reached. When this explanation is shown to the original expert it is often found that he or she agrees with it or at the very least that there are no objections to the suggested strategy. This means that an explanation has been easily attained about an area which may otherwise have proved to be problematic.

For example, one of the experts used to build the bridge system found it hard to explain the reason for choosing a certain type of end support, that is whether to use either an abutment or an bank seat. When another expert was asked about this subject, he said that the choice largely depended on the width of the bridge. Therefore a rule was elicited without the original expert having to struggle to provide an explanation.

Many experts find it difficult to identify certain parts of their expertise (Ericsson and Simon, 1980; Hartley, 1982), and in these situations an alternative source of explanation is useful.

An alternative source is especially useful when dealing with design, which generally involves a large number of heuristics that experts often find difficult to verbalise.

Even if the knowledge engineer is very familiar with the domain and the terminology used within it, there will still be times when the principles described are unclear. Using more than one expert means that an individual does not have to explain repeatedly the same domain concept to ensure that the knowledge engineer understands. Another expert can be used to help explain. This helps to avoid the expert becoming frustrated with the inability of the knowledge engineer to understand.

Thus, alternative sources for explanation help to overcome the "Grandma" or "black box" effect, where the expert feels that they have to oversimplify the explanation to enable the knowledge engineer to understand. This over-simplification can lead to misrepresentation of knowledge and loss of important information (Waterman and Jenkins, 1979; Fox *et al.*, 1983).

Two Different Approaches to the Domain Available

Experts inevitably adopt differing thought mechanisms, and frequently they will also adopt different approaches to problem solving. These differing approaches can help to create a more detailed picture of the domain, as more possible decision routes are considered.

This is especially important when building design systems, as design is subjective and consequently the more possible decision routes which can be identified, the more complete the resultant system will be.

Despite the varying approaches used by experts, it is frequently found that the same solutions are finally obtained. Therefore, the differing approaches help to assess the solution paths employed. This then enables the most efficient route to be identified.

Less Chance of Missing Vital Information

Despite taking the utmost care with KE, the Knowledge Engineer will almost inevitably fail to unearth some vital information. Multiple experts help to ensure that this problem is less common, as because there are more experts being used, inevitably they will cover different ground, helping to ensure that more of the entire domain is covered.

Elicitation of the Exceptions to the Rule

On a similar vein to the previous section, more than one expert can help to identify "exceptions to the rule", that is the situations to which the usual rules do not apply. This is a recognised problem in KE (McDermott, 1980), particularly in design, where the exceptions can be numerous. Often, it is these exceptions which are the most interesting area in the domain as they help to illustrate the expertise, so it is important to include as many of them as possible in the KBS. The inclusion of exceptions to the rule can help to make the difference between a KBS which is acceptable and a KBS which is good at its job. Often, a KBS will fail because it cannot cope with an "exception" to the rules it contains.

Preferences Owing to Familiarity Counterbalanced

Inevitably, experts are going to be biased towards the domain concepts with which they are most familiar. Alternative experts can help to counteract this bias. This is especially important as design is so subjective.

Bias is discussed in more detail later in this chapter and also in Chapter 2.

Reduction in Time

By using more than one expert, generally the amount of time spent interviewing will be increased. However, the overall duration of a project can be reduced as the inaccessibility of the expert is drastically reduced, so far less time is wasted waiting for an appointment. Also, the amount and accuracy of the information elicited is increased, so less time is needed to improve the developed prototype, as the knowledge base is more complete than if one expert had been used (see the chapter on evaluation).

This reduction in time can be important, especially when the expert needs to see a result in order to sustain enthusiasm for the project. By using more than one expert, progress is apparently rapid owing to the knowledge which has been acquired from the other experts. Therefore, the knowledge base appears to grow very quickly and the expert does not feel as if the KE is too arduous. Also, an individual expert has to commit less time to the project as he/she does not have to supply the entire knowledge base since knowledge is acquired from other sources. This helps to maintain expert enthusiasm, and this enthusiasm helps to stimulate the expert into providing more knowledge.

Immediate Verification Procedure

Feedback to the experts is a very important part of KE, and this point is discussed later in this chapter. If rapid prototyping is not being used, summaries of the knowledge elicited and used to date should be sent to the expert between interviews, so that they can assess the progress which is being made. One of the biggest benefits of using more than one expert for KE is that, by reviewing this feedback, the experts act as a verification procedure for one another, checking the information which has been elicited from other experts. If the same rules or ideas are supplied by more than one expert, the knowledge has been somewhat validated. If different solutions are suggested, further analysis of this domain concept is required, so the differences can be accounted for.

Obviously this rough form of preliminary checking is not sufficient to ensure the completeness and accuracy of the knowledge base. A more complete evaluation process is necessary to ensure the correctness and applicability of system. System evaluation is treated as a separate topic and is discussed in detail in Chapter 7.

Disadvantages of Using Multiple Experts

Inevitably there are also some disadvantages associated with using more than one expert for the KE. These are now discussed.

Possibility of Clashes of Information

The problem of conflicting information has previously been seen as the major disadvantage of multiple experts for KE (Greenwell, 1987).

In our experience, conflicting information has not been a difficult or regularly occurring problem. We have found that usually experts agree on the major domain concepts, any disagreements which do arise proving to be minor and relatively easy to overcome.

It is recognised that if conflicting opinions had arisen, they would have proved to be problematic, as there is obviously a risk of producing a system which is unreliable owing to the clashes of information within the knowledge base.

There are ways to combat this problem. One is to expose the conflicting viewpoint to the experts and ask them to suggest a compromise or to explain the reasons for the difference. If the experts agree that the alternative approaches are viable, then the problem becomes irrelevant as one or both of the approaches can be included and the reasons for the discrepancy explained. If the experts do not agree on the feasibility of both approaches, an explanation as to why they see it as inappropriate can help to counter the disagreement, again resulting in the information being successfully combined.

An alternative approach to combining conflicting information is to use a dominant knowledge source – that is, an expert who is chosen as the "main" knowledge source and whose decision is taken to be over-riding. Any disagreements which do arise can then be decided by this dominant expert. The dominant expert should be chosen at the start of the KE. The choice can be based on a number of factors, including availability, how articulate the expert is, enthusiasm and adaptability. Basically, choose the expert to suit the task.

Questioning of Integrity

Another possible problem with using more than one expert for the KE is that the experts may feel that their integrity or expertise is being questioned.

If the experts are senior and secure in their status as "experts", this should not be a problem. However, in situations where the experts used are not as confident about their expertise, they may feel their ability is being questioned. In extreme cases they may see the other experts as a threat to their job. It is important that the knowledge engineer stresses the importance of their co-operation, explaining to each of them that the use of additional experts is to reduce inaccessibility and to improve the knowledge base, and it is not questioning their ability.

Covering the Same Ground Twice

Inevitably, by using multiple experts, much of the same ground is covered by more than one expert. The time lost by this repetition, however, is thought to be very small and more than compensated by the overall saving in project time, which has been discussed earlier.

The Importance of Assessing How Your Experts Think

The way an expert thinks inevitably affects his/her reasoning patterns, and so it should influence the style of interview adopted. Careful choice of interviewing technique is vital in KE. The knowledge engineer has to try and assess which methods will best suit a given expert. If unsuitable techniques are chosen, this can greatly affect the productiveness of the KE and may even cause the expert to lose interest in the KE.

Designers tend to think pictorially: ask a designer a question and he/she will invariably reach for a pen and paper, either to help with the understanding of the question, or to clarify the answer. Removing this pen and paper is like limiting the number of words they can use. Therefore, the KE should be oriented towards using visual techniques, which suit this style of thinking, employing a variety of pictures, flowcharts and graphical images.

Individual styles of thinking also need to be considered. We have encountered numerous thinking styles through our experience of KE with different people: for instance, one expert we have dealt with is a very methodical thinker, preferring to describe every detail and consequence of a decision. Another expert who was used in parallel to this methodical expert relied on a totally different spontaneous style of thinking. Taking these differing thought mechanisms into account is very important, as obviously interviewing the above two experts in the same way would not be very productive. The way in which an expert thinks can usually be assessed by watching the way in which they act, speak, answer questions and deal with problems (see the section on Unstructured Interviewing later in this chapter, which recommends the "shut up and listen" technique). The way in which a person thinks is often reflected in many facets of personality. Body language can also reveal a lot about the attitude to you and the problem: for instance, generally when people fidget while they are speaking, it indicates that they are unhappy with the response they are giving (unless they are the type of person who fidgets all of the time!). Pauses can also reveal when a person is finding something difficult to explain: beware, they may use the easy option instead of delving into their memory to find a suitable explanation (see the section on Alternative Source for Explanations, earlier in this chapter, for the "Grandma" effect).

In order to perceive the way in which a person thinks, a certain amount of intuition and receptiveness is required on the part of the knowledge engineer, but we feel that this is true for nearly all aspects of knowledge engineering! Be warned, however: we all make mistakes. If you have chosen a KE approach, but the expert does not seem to either like it or feel comfortable with it, be prepared to change your mind and use an alternative. With practice, however, you will become better at choosing techniques with which the expert is comfortable.

If there is more than one knowledge engineer (and incidentally this can be very beneficial as one can take notes, watch the expert etc. while the other asks the questions) take note of which of the knowledge engineers the experts reacts better to. Inevitably, the expert is going to develop a a better rapport with one than the other, and this knowledge engineer should dominate the questioning.

The Factors Which Help You to Assess the Experts

As discussed earlier, finding suitable experts is an important part of successful KE. Once a suitable expert or experts have been found, the next step is to try to assess the way in which they think and the way in which they deal with the problems they encounter. By doing this, suitable forms of interview which complement these approaches can be chosen. This assessment can cover a number of factors.

Co-operation

Firstly, how co-operative are your experts? Are they only co-operating because their boss told them to? Or are they only helping as a favour to you or your superiors? If an expert is not inclined to co-operate it can make the KE unnecessarily difficult and it can even affect the knowledge base which is developed. Time and care must be taken to promote their interest and stress their importance in the project.

Enthusiasm

If the experts are enthusiastic about the project, the KE is going to be made much more easily. If the KBS is to be of direct use to the experts themselves, then stimulating their enthusiasm should not be difficult. If, however, as is often the case, the system will not directly benefit the experts but only their subordinates, they may actually find it difficult to see what the advantages of developing the system are. In this case, the benefits of the system to the company must be made clear and any indirect advantages emphasised.

Articulateness

Another factor which must be considered when assessing experts is how articulate are they? As mentioned earlier, many experts find certain aspects of their expertise difficult to verbalise. Try to use interviewing techniques which help identify the experts' thought patterns or which help them explain the mechanisms they are using, such as pictorial methods or methods which map out the solution paths being followed. It is very tempting to try to help the experts explain their meaning. This is not recommended, as your intervention can influence the expert and result in a tainted view of the expertise. Try and keep quiet and let the experts work it out by themselves!

Bias

As discussed earlier, bias is an inevitable problem with KE. For instance, when developing the KBS for conceptual bridge design, one of the experts used was strongly biased towards steel bridges. This had to be taken into account when building the knowledge base, to ensure that a more objective system was developed. Bias is present in some form, no matter how minor, in most experts. Therefore, it is very important to recognise its presence whenever possible and try to counteract it.

Finding out about experts' preferences from their colleagues can help to identify any biases which may be present: but take care to choose colleagues who actually respect them and their work! Otherwise the problem of bias will still be apparent.

Confidence

The expert's confidence in his/her own expertise must be taken into account. We have been fortunate that the majority of experts we have dealt with have been secure in their expertise and position. However, this is not always the case. Lack of confidence can affect the information being elicited, so once it has been established that their expertise is genuine, try to assuage this insecurity wherever possible.

How Comfortable is Your Expert?

This does not just mean comfort in the physical sense, although that is also important! If the expert prefers to be on home ground, then the interview should be carried out in his/her office. If, however, when in his/her own office he/she seems to be continually distracted, then maybe an alternative venue would prove to be more productive. If the interview is being carried out somewhere which is unfamiliar to the expert, try to make the environment as comfortable as possible, so the expert feels happy with the surroundings. This can help him/her to feel at ease with the KE process, which increases the productiveness of the interviews.

Also, is the expert comfortable with the KE techniques which you are using? The importance of this is stressed in the following sections, and advice on how to choose suitable interviewing techniques is given. This point is possibly the most important and interesting factor in successful KE: recognising the expert's thought patterns and trying to adjust the KE to suit.

Choosing an Overall Approach to KE

KE should be reasonably structured if it is to succeed, so an overall strategy must be determined early in the project. Generally, KE can be split into two stages: unstructured or preliminary interviewing and structured interviewing. The preliminary interviewing stage is required regardless of the approach adopted. The structured interviewing stage will vary depending on the chosen approach. It is usually one of two types: rapid prototyping or sustained interviewing.

Rapid prototyping builds the first prototype using the limited information which has been elicited during the preliminary interviews with the experts. Criticism of this first prototype is then used to elicit the remainder of the knowledge, with a large number of subsequent prototypes being developed which incorporate these criticisms.

The alternative approach is to use sustained interviewing. In this case, the majority of the domain knowledge is elicited before the first relatively accurate prototype is built.

The choice of approach will depend on the domain and the experts being dealt with. In our experience, however, for design oriented domains, rapid prototyping is usually inappropriate.

Is Rapid Prototyping Suitable?

When we started to develop our first design-based KBS, advice was followed (Hayes-Roth *et al.*, 1983) and a prototype was attempted once the preliminary interviews had been carried out, but it was found that this prototype was virtually impossible to build.

The main reason for this was that the chosen domain was design oriented, very large, fairly case specific and contained a very number of unusual cases and exceptions to the rule. The information which had been elicited from the preliminary interviews was sparse and therefore difficult to combine. Links between the disjoint information elicited had to be guessed, resulting in gross inaccuracies.

If a weak prototype is presented to the experts, even in rapid prototyping, there is a danger of losing their confidence in the project. In addition, much design knowledge tends to be case specific, so it is difficult to build a system by criticising a prototype, making rapid prototyping inappropriate.

There are other considerations when deciding which approach to use. For instance, how computer literate are the experts? Most engineers are familiar with computers and their uses. However, a large number of experts are senior engineers who may be familiar with the applicability of computers, but who are probably not as confident using them, and it is unlikely that they personally will regularly use computers. As has already been mentioned, interviewing techniques should be chosen to make the experts feel comfortable. Therefore, is it sensible to use computers so prominently in the KE?

Rapid prototyping is, however, a very good KE technique. It is quick, efficient and, if the experts are computer literate, it can increase their interest and enthusiasm. It also creates systems which have already been partially evaluated. In other domains, rapid prototyping is an approach which should be given careful consideration. But if it does not seem to be suited to your domain, do not be afraid of going against the standard opinion by using another approach to KE.

If rapid prototyping is not adopted then alternative interviewing techniques must be used during the structured interviewing stage. These differing techniques are needed for a number of reasons. Firstly, choosing the technique to suit the expert or situation is very important. Secondly, varying the type of interview helps to prevent the experts becoming bored. Also, by varying the techniques, different information can be elicited, as the domain is approached in a number of ways. Finally, without rapid prototyping, some form of feedback must be incorporated into the KE as an indication of the progress being made.

Interviewing Techniques
Unstructured Interviewing

As mentioned earlier, an unstructured interviewing stage is needed regardless of the KE methodology chosen. This stage aims to assess the target domain, identify the major domain concepts and establish a rapport between the experts and the knowledge engineer. The importance of establishing this rapport has been stressed elsewhere (Breuker and Wielinga, 1987; Gammack, 1987; Welbank, 1987). In our experience, this is a very important part of the KE process, helping to make it more productive, and more importantly, more enjoyable for both the knowledge engineer and the experts. Where possible, it is good to get to know your experts and often you will find that you actually like them and they get to like you. Show an interest in what they do and don't just jump straight into the KE: a little courtesy goes a long way! This obviously greatly helps the KE, as the barriers which exist between strangers will be dropped.

The unstructured interviewing stage should allow the experts to talk freely and without interruptions on their subject. It is important that the discussion is uninterrupted as otherwise the expert can be distracted, causing the interview to lose direction and continuity (Ellman, 1987; Welbank, 1987). This "shut up and listen" technique is not as easy as it sounds, but it is very productive and is good for establishing the domain.

The unstructured stage should also be used to assess the experts, helping to identify KE techniques with which they will be comfortable, the importance of which has already been discussed. It also helps to identify whether the experts are suited to KE, or whether they fall into one of the "unsuitable" categories which have already been listed.

Structured Interviewing

As discussed earlier, rapid prototyping is an approach to KE which relies on the repeated review of successive prototypes in order to acquire the necessary expertise. The first prototype is built using the sparse knowledge which has been elicited during the preliminary interviews. It is hoped that by criticising a partial prototype, and by identifying the gaps and errors in this prototype, the necessary expertise will be gradually elicited. The advantages and disadvantages of rapid prototyping as we see them have already been discussed.

If rapid prototyping is not used, alternative interviewing techniques must be chosen and adopted in order to sustain the experts' interest and to ensure that different sorts of knowledge are elicited.

Alternative Interviewing Techniques

These interviews are intended to verify and explore the domain concepts identified during the unstructured interviewing stage. The domain concepts are the main ideas which constitute the domain.

Pictorial Representations

This technique relies on the selection of pictures which cover an area of a domain. The selection should cover redundant options and options which are known to be inappropriate or outside the domain concept, that is the section of the domain which you wish to cover as well as those which are felt to be viable. These pictures are shown to the experts and they are asked to comment on them: saying which options they would use and why the others would be inappropriate. The pictures act as triggers, enabling the expert to comment more easily on the domain.

One advantage of pictorial representations is that they can help to prevent experts digressing from the subject area being assessed, which is a common problem with experts. They can also remind experts of specific examples which they have dealt with in the past and this can help to focus experts who habitually talk in general terms as well as helping the all important "exceptions to the rule" to be elicited.

A particular benefit of this approach is that designers like pictures: as mentioned earlier they tend to think pictorially, so this is a very sympathetic form of KE and therefore it tends to be successful, as designers are comfortable with the approach.

We have used this approach to elicit information on the end support choice for the bridge system. This was an area which proved to be particularly problematic as the experts found it difficult to identify end support options which they would not use. The pictures which were used greatly helped to clarify this (Fig. 5.1).

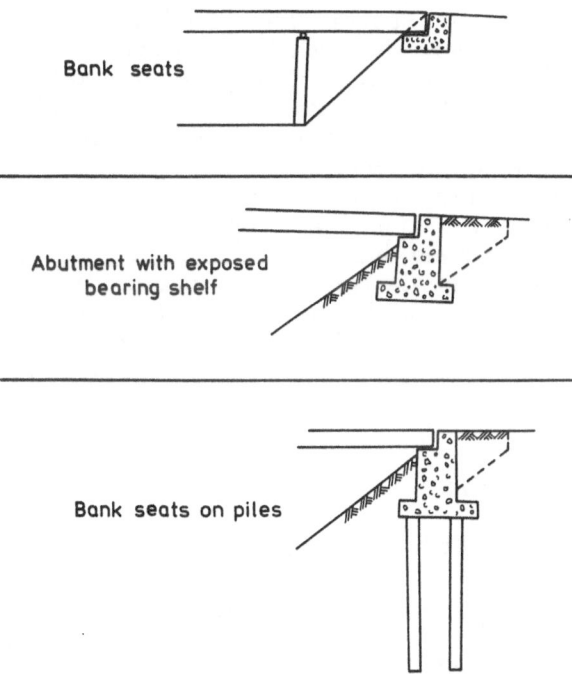

Fig. 5.1 Pictorial representations

Card Sorts

This is a technique which is particularly useful for establishing links between domain concepts.

In this technique, all the major domain concepts and relevant elements belonging to the domain (or the area of the domain which you want to concentrate on) are written on cards. Blank cards are provided so that any missing elements can be added by the expert. The expert is asked to arrange the cards in any way which seems appropriate, adding in extra cards where necessary. This arrangement is then photographed and the reasoning behind the arrangement discussed. In this way, the way in which the expert has approached the domain can be easily assessed. The expert is then asked to rearrange the cards in a different way. Again the arrangement is photographed and the expert is asked to explain the new links between the cards or card groups. This process is repeated three or four times until a number of different arrangements have been considered (Fig. 5.2). These varying approaches can then be compared, which helps to reveal the way in which the experts would deal with the problems which constitute the domain in question.

Card sorts help to clarify the domain concepts in the minds of the experts and the knowledge engineers. The approach helps to ensure that all the concepts within the domain have been elicited, forcing the experts to consider the domain concepts and relationships in different ways. It is an approach which forces experts to consider the domain and the relationships within it in a number of ways, often including approaches which they would never normally use. Consequently, it is a good way of eliciting information on subjects they would not normally consciously consider. It is an approach which helps to make the links which are present within the domain clear. Card sorts can also help to clarify the overall approach used by the expert to the problem domain.

In our experience, all the experts who have used this exercise have found it challenging and enjoyable, as it is a novel approach which stretches their expertise. Consequently, they were willing to commit more time and effort to it. This approach is particularly suited to an intuitive thinker as it allows rapid decisions and is quick to do. However, it will only work with experts who are willing to try new ideas. Experts who are very sceptical of the approach before they start will generally not take the approach seriously and will not give the arrangements the consideration they deserve, and consequently the interview will tend to be unproductive.

Repertory Grids/Triples

This is a technique (Kelly, 1955) which we have not actually used, so we cannot give actual examples of its success. However, other knowledge engineers have used it and found it to be very effective (Shadbolt, 1986; Greenwell, 1987).

In this technique, the expert is asked to list a number of significant objects or elements from within the domain (usually between ten and twen-

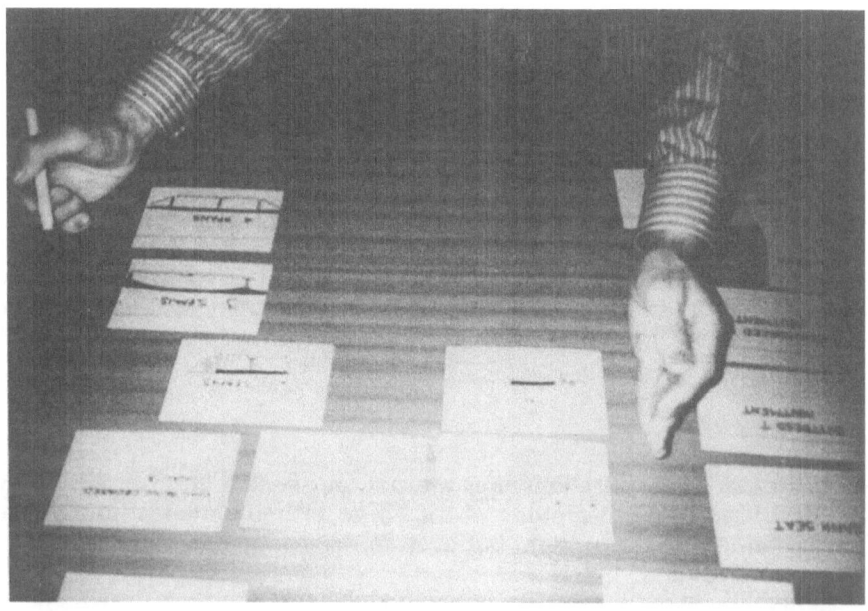

Fig. 5.2 Card sorts

ty). Three objects are randomly drawn from this list and the expert is asked to identify the two which are most alike and the one which is most dissimilar, stating the reasons why this is so. This process is repeated until a reasonable number of relationships or "constructs" have been elicited. Then each object is rated on a scale for each construct, to form a "grid" of results. A statistical analysis can then be used to compare the objects in the grid.

There are a number of variations on this basic technique (Hart, 1986; Shaw and Gaines, 1988; Greenwell, 1987). Probably the most important use of repertory grids is to elicit the data for the structured interviews (Greenwell, 1987). They are also good for helping inarticulate experts, as the expert can make lists of the domain concepts which can be manipulated instead of having to talk about the domain as an entirety, and most people find it easier to list and link subjects than to discuss them coherently.

Questionnaires

In our experience, questionnaires are not an effective form of KE. They are best used at the beginning of the KE to promote interest and gauge the usefulness of the system, but even in these situations their effectiveness is doubtful.

Generally, experts are very busy people and they are unwilling to spend time on something which they feel is of little benefit, such as a questionnaire. Filling in a questionnaire is time-consuming and the amount of information elicited from them compared to that acquired during a meeting of much shorter duration is very small.

There is also a certain apathy associated with filling in forms. Often brief answers will be given (if they are given at all!) when in fact more description is required. If the same questions were asked verbally, such abbreviation would be less likely, and even if brief answers were given the knowledge engineer could immediately ask for further explanation. Interviews involve direct contact with the experts, so they tend to be more interesting and the experts' answers will usually be given more consideration.

Case Studies

In design oriented domains, case studies are inevitably very important in the KE, and no KBS should be considered complete unless a number of case studies have been used in its development.

However, it is often found that the case-specific nature of the information elicited using case studies does not warrant the amount of time involved. Frequently, a very large number of case studies are required to cover a very small part of the design domain. Therefore, it may be better to concentrate on using less focused techniques, only using actual case studies during the later stages of KE and the system evaluation. At this stage, the case studies can help to identify exceptions to the rule or areas which have been missed during the early KE.

Paper Models

Paper models are like large diagrams which depict the knowledge and relationships elicited (Fig. 5.3). They can be used to describe certain domain concepts or to present the knowledge engineer's impression of the expert's overall approach to a problem. Effectively, they are paper versions of the eventual system, and therefore the apparent benefits are similar to those found when using rapid prototyping without the programming or computational difficulties.

The experts are asked to analyse the paper model, criticise and comment on it, checking that the knowledge engineer's interpretation of their expertise is correct.

Paper models provide a good representation of the knowledge acquired and therefore are a good form of feedback. The knowledge contained acts as triggers, so the model is also good for eliciting new information. The structure of a good model should be easy to understand and analyse, and forces the expert to go through the information methodically, so it is a good review structure. Paper models are good for revealing gaps in the knowledge base and, when an aspect of the model is found to be incorrect, replacement knowledge has to be provided by the expert, otherwise an obvious gap in the model is created. The expert rejecting knowledge and not providing alternative information is a problem which can be faced with other interviewing techniques.

The main disadvantage of paper models, both for the experts and the knowledge engineers, is that they are time-consuming and demand a high level of concentration. They are only really suited to methodical experts who are comfortable with considering large amounts of information for long periods of time.

In terms of preparation time, however, usually a knowledge engineer will draft a paper model before encoding the system anyway, so there is no extra time incurred when using this technique for the KE.

Brainstorming

This is a KE technique which is useful if more than one expert is available. The experts are put together to discuss certain domain concepts. We have never actually used this approach, but we would have if we could! Unfortunately, in our experience it is hard enough to get an interview with one expert at a time, let alone two! (See the section on Inaccessibility.) However, as mentioned earlier, this is a good approach which can help to settle disagreements, as well as elicit new information. It can also help to extend the information which has been elicited, as it is likely that the experts will become interested in the discussion and expand to new areas which may have never been covered.

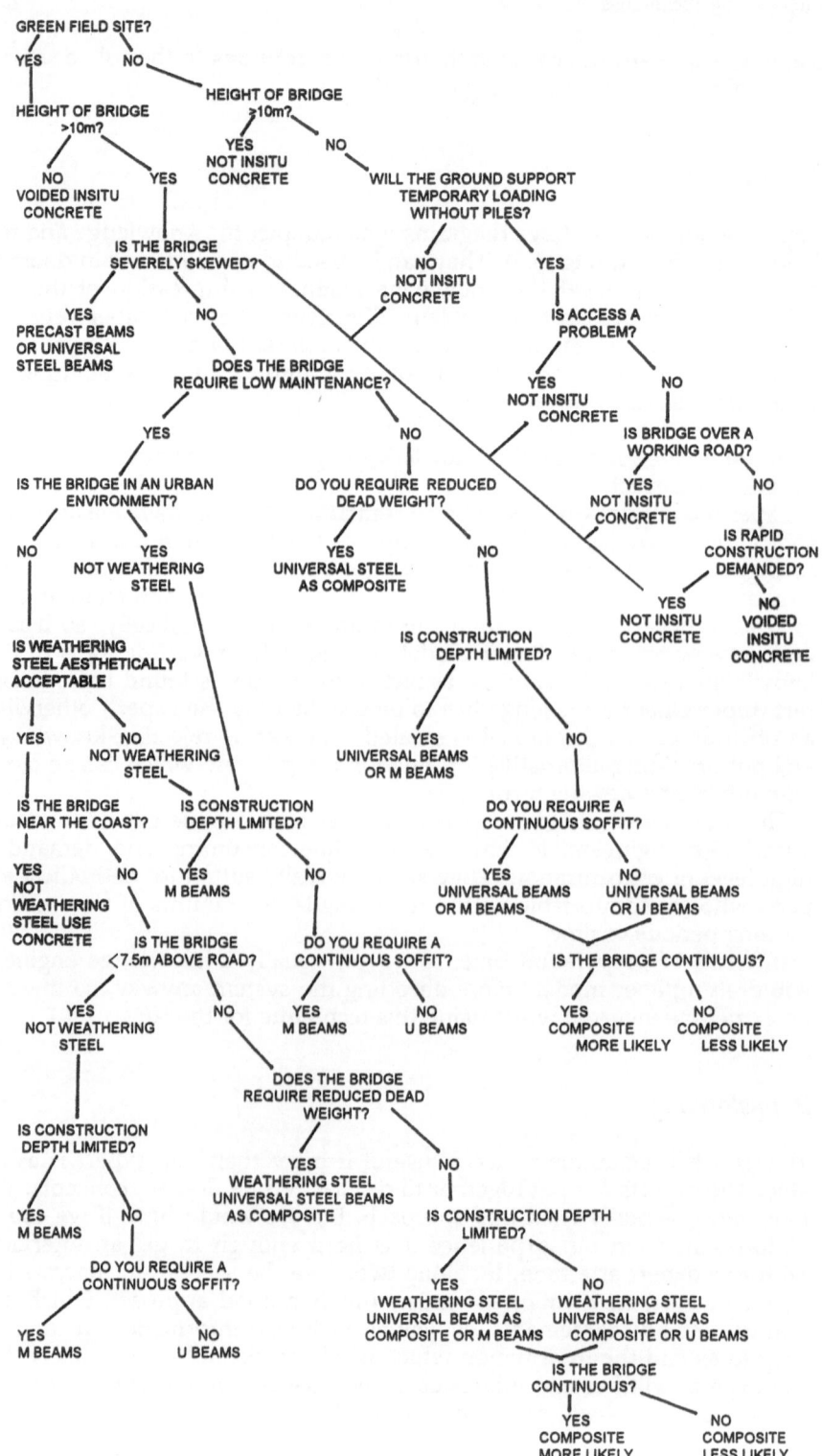

Fig. 5.3 Paper models

Role Playing

Role playing can take two forms. In the first case, the knowledge engineer takes the role of a client and the expert provides the solutions, enabling the techniques he or she is using to be analysed. Alternatively, the knowledge engineer plays the expert so that the real expert can review the progress and check that the knowledge engineer has gained a clear impression of his/her expertise. Inevitably, this latter case can only take place at quite an advanced stage of the KE. This can be very effective but care must be taken to ensure that the expert does not feel that his/her integrity is being threatened.

Again, this is a technique which we have not actually used. This is really owing to the types of experts which we are used to dealing with. Many engineers would not feel comfortable with this approach: certainly the experts we have dealt with would have felt self-conscious "play acting" and carrying out such an unfamiliar role. Again, the knowledge engineers must use their judgement to decide if this is an approach with which the expert will feel comfortable.

Choosing Interviewing Techniques

When trying to choose suitable interviewing techniques, the best advice is to look carefully at your expert(s): assess what they feel comfortable with, the way in which they think about the domain, the techniques they use to help them explain things, and choose the interviewing techniques accordingly. For instance, it has already been stated that pictorial methods suit designers as they tend to think pictorially. Similarly, it is not sensible to force spontaneous thinkers to analyse a paper model, as it would not suit their style of thinking. A form of interviewing which allows a more rapid style of thinking (e.g. case studies) is more appropriate.

Make sure you monitor the success of the interviewing techniques which you use, so that inappropriate techniques can be disregarded and successful ones repeated.

There are other considerations when choosing a suitable interviewing technique. For instance, using a paper model can be very productive with the right expert, but it is only feasible if quite a long time is available. Often this will not be the case, so shorter, less time-consuming approaches should be adopted. Similarly, it is no good deciding to devote an entire interview to the analysis of case studies if there are only a small number of case studies readily available. It would be better to keep these for a later stage in the system development, when they can be used to help evaluate the system (see Chapter 8 on Evaluation).

As well as checking the availability of the necessary resources, always forewarn the expert of the structure and content of the forthcoming interview, so he/she can prepare for it. This helps to improve both the efficiency and productiveness of the interview.

Finally, it helps if the experts are enjoying themselves. Techniques such as card sorts and repertory grids are fun to do and they stretch the experts by making them consider the domain in a new way. In our experience, testing the experts in this way can be particularly productive as it increases their interest, so they are prepared to commit more time and energy to the interview.

Recording and Interpreting the Knowledge

A detailed and careful approach to interpreting the elicited knowledge is needed, regardless of the KE methodology adopted. Whether the knowledge has been elicited by interviewing techniques or by criticism of a prototype, the information must be interpreted and encoded.

Whenever possible, the analysis of the elicited knowledge should be carried out by the knowledge engineers themselves. After all, a lot of information could be missed by someone who is unfamiliar with the target domain or indeed the experts!

There are a number of ways in which the analysis of the knowledge can be made more efficient and productive, some of which are discussed below.

Transcript Analysis

Every interview should be tape recorded, preferably using more than one tape recorder, as unfortunately tape recorders have a habit of failing at crucial moments! The experts should always be consulted on this matter, but in our experience, most experts are familiar with recording information and so are happy for tape recorders to be used.

Tape recording an interview provides a permanent record of the interview, which is vital as it is unlikely that a knowledge engineer will be able either to note down or remember all that has been said. It also allows the knowledge engineer to concentrate on the interview, and not be distracted by taking notes. The only notes which then need be taken are those relating to the expert's body language and pauses, about the tools which he/she is using and so on. Be careful, though – some experts feel intimidated when knowledge engineers take notes while the interview is being tape recorded: it makes them want to know what is being written!

Every interview should be transcribed in full. Although this is a long and very tedious process (transcribing an hour long interview can take three to six hours!), full transcripts allow the interview to be accurately analysed. It also helps to ensure that the knowledge engineer begins to get a good understanding of the domain, as he or she becomes familiar with the terminology and concepts put forward: this helps to ease the KE as the experts does not have to explain every technical expression.

When first reading the transcript, highlight sentences, statements and key words which are of interest. Each transcript should be thoroughly read at least twice to ensure that all the relevant information and triggers have been identified. Whenever possible, try to get someone else to read the transcript through so that they can spot anything you might have missed. Detailed analysis at this stage can help to create a more complete knowledge base and consequently save a lot of time later on!

Always try to analyse the transcripts as soon after the interview as possible, while it is still fresh in the memory, so additional notes about the style of the interview can be made, such as notes on whether there were any interruptions, and the effect of these interruptions on the expert's train of thought; the "props" the experts used to help explain things; whether he/she was sketching anything at the time and so on. These notes can act as good clues when interpreting the interview. If the expert did use any "props" during the interview, where possible, try to obtain them so that they can be used in the analysis of the interview. Similarly, as is often the case with designers, if the experts make rough sketches during the interview to help them explain certain ideas, always make sure that you take these rough sketches with you at the end of the interview: they can prove to be invaluable!

When analysing the interview, as well as identifying the rules and heuristics contained in the transcript, try to look at the manner in which statements are made, what the experts are not saying and what they are apparently side-stepping. A great deal can be learnt from this hidden information!

Analysing the transcripts can give a good indication of what the next interview needs to focus on, as they are a good record of the areas which have already been covered with each expert. They also enable the interviews to be compared.

There are computer software packages now available for analysing transcripts. They can be programmed to pick up common words and phrases, and they can also be programmed to recognise key words in the interview. This helps to identify automatically the important parts of the transcript. However, we have not actually used such a package, so we cannot report on their effectiveness.

In a similar vein to tape recording the interviews, videoing the experts during interviews and during case studies can be useful. It has the advantages that the interviews can be watched repeatedly as well as listened to, which allows the experts' body language, expressions and the tools which they use to assist them to be assessed. All of these obviously help in the appraisal of the interview. However, there are problems associated with this approach: primarily that many experts would not feel comfortable being videoed. A tape recorder is far less conspicuous than a video recorder! As mentioned earlier, it is very important that the experts feel at ease if the interview is to be successful. It is up to the knowledge engineer to assess whether the experts would feel happy being videoed. There is the additional problem that videoed interviews inevitably take longer to set up. Many experts would become frustrated with this apparent "waste" of time. Alternatively, the experts can be brought to a room where the video etc. are already set up. This, however, has the added problem that strange surroundings (particularly with a video cameras present) can cause the expert to feel uncomfortable, which would impinge on the interview. Again, it is a question of assessing your experts and deciding whether videoing is a suitable approach.

Intermediate Representation Techniques

Intermediate representation techniques are methods which represent the elicited knowledge in forms that help to bridge the gap between the transcript and the encoded form of the knowledge. Techniques are used which are nearer to the encoded version of the system but which are not too complex or computer oriented. Representing the elicited knowledge in an intermediate form can increase the effectiveness of the knowledge analysis, as it allows both the knowledge engineer and the experts to review the elicited information and the way that it has been interpreted in a simple form that is closer to the encoded version. This verifies the elicited knowledge and identifies areas in need of further analysis, as well as helping to ensure that nothing is lost in translation form interview to computer. Intermediate representation also eases the transition to the chosen form of knowledge representation and consequently the encoding of the system.

The form of intermediate representation chosen is influenced by a number of factors. Primarily it is influenced by the type of knowledge representation adopted and by the chosen development software.

Many of the shells which are available provide an easy to read and understand rule or frame format which somewhat replaces the need for intermediate representation. Care must be taken, however, that the relationships between the rules or within the frame hierarchy are well understood. An alternative form of intermediate representation can help to establish this. When the implementation tool or language chosen has a high degree of representative power, then an intermediate representation which only outlines the main ideas should suffice.

If a language or a shell which uses a more complex form of knowledge representation is to be used, then some intermediate representation could well ease the task of encoding.

Knowledge representation, the techniques which are available and when they can be suitably used, are discussed in detail in Chapter 6. Inevitably, the type of knowledge representation adopted is going to influence the choice of intermediate representation.

Forms of Intermediate Representation

One form of intermediate representation is the paper model, which has already been described. The benefits of paper models are that they are easy and therefore quick to develop and understand, and they are an efficient form of feedback as well as a good form of intermediate representation. We have found that designers like paper models as they are familiar with following through a systematic flow of information. If a rule-based approach to programming is to be used, paper models are particularly useful.

If an object oriented approach to programming is to be adopted, an alternative intermediate representation technique may be preferable. Using such an approach to programming can make the representation of some domains much simpler, as is discussed in Chapter 6. One intermediate representation suited to the description of objects is the epistemic net (Keravnou and Johnson, 1986). The epistemic net is used to define a concept or classify and relate entities into a logical structure. The basic principle behind an epistemic net is selecting an object, this selection being based on one or more pre-defined conditions. The net relates the object's distinct characteristics under the same heading, in much the same way as a frame. A status transition diagram is then used to describe each object's "status", the transitions which it can make and the conditions which control these transitions. This information is then represented as a network of object/status nodes with permissible conditions attached.

Relational networks (Keravnou and Johnson, 1986) are another form of intermediate representation for objects. They depict the mappings between factual entities, thus describing how objects relate to one another. The completed network describes a concept in terms of relevant entities (Greenwell, 1987).

Unfortunately, many of the methods for intermediate representation which have been suggested have not been tested on real domains (e.g. Brouwer-Janse and Pitt's, 1986 technique, which structures knowledge by directly analysing the problem-solving behaviour of the experts).

Basically, there are no restrictions on other methods which can be used for intermediate representation. All intermediate representation should do is present the knowledge in a form which eases the transition to the final knowledge representation format chosen and hence the encoding of the system.

Intermediate representation can be used solely to help the knowledge engineer interpret the knowledge and/or as a form of feedback to the expert.

Feedback

When rapid prototyping is not being used, feedback is an essential part of the KE. The expert must be made aware of the progress being made, so he/she sees that the effort is worthwhile. Without this feedback it is unlikely that the expert's enthusiasm will survive. Feedback is also good for maintaining good relations with the experts and assessing the effectiveness of the KE approaches adopted.

Following each interview, once the knowledge has been analysed, a summary of the information elicited should be sent to the experts. This is beneficial in three ways: firstly, it is a progress report; secondly, it acts as an initial verification procedure, allowing the expert to evaluate your interpretation of his/her knowledge; thirdly, it provides a record for the expert to refer to before the next interview.

Some of the KE techniques are good feedback techniques in themselves, such as the paper model and role playing techniques, as both these allow the expert to check the elicited knowledge.

Inevitably, if rapid prototyping has been adopted then feedback is not a problem, as each prototype should provide sufficient feedback.

Conclusions

In conclusion, it must be stated that KE is not as easy as it initially appears. We are some of the many who have assumed it was merely a task of asking questions, and who have found that this is not the case. KE must be taken seriously if it is to succeed: suitable experts must be found, these experts must then be assessed and interviewing techniques must be selected to suit them, and then each interview must be transcribed and carefully analysed. It takes time, patience and organisation to elicit a very small portion of a person's expertise.

It is very important to be flexible if the KE is to be successful. Be prepared to see your experts whenever they want, change approaches if they are not working or the experts seems uncomfortable with the interview, and be prepared to alter your sights if they do not seem realistic. Finally, be warned: standard approaches may not suit your needs. Look at the people, the type of domain and the surroundings you are dealing with, and choose your approaches accordingly.

Chapter 6

Knowledge Representation and Control

Objectives

This chapter aims to:

- Briefly define AI and knowledge representation.
- Identify the choice of tools with which KBS can be built.
- Investigate various ways of representing knowledge.
- Discuss different search and control mechanisms.

Introduction

The concept of Knowledge Representation (KR) is basically easy to understand: it is the process of writing down, in some form, descriptions of a domain or some state of that domain so that it can be manipulated to form new conclusions. In Artificial Intelligence work, the main concern is to represent knowledge in a way which enables a computer to assimilate and manipulate it effectively.

In a sense, every computer program contains knowledge about the problem it is solving. A characteristic of conventional computer programs is that the knowledge contained is not expressed explicitly and cannot be readily expanded or manipulated (Waterman, 1986).

Conversely, two of the desirable features of current KBS are transparency and adaptability. This is partly why effective KR is so important in KBS development, but it is also because KBS tend to deal with some aspect of human intelligence.

The demands of AI systems consequently differ from those of standard computer programs; so a brief definition of AI is included in this chapter, with a description of the different set of interests which are apparent when developing KBS.

There is unfortunately still neither a perfect nor a standard method of KR. This stems largely from a lack of understanding of what knowledge actually is. Knowledge, what it is and how it is manipulated, is inevitably a very important criterion in KR. There is still some uncertainty about the way people think, but some of the factors which have been established and which affect KR are identified in this chapter.

Finally, the way in which knowledge is used is equally important as the way in which it is represented. A number of control and search mechanisms and their relative merits are therefore discussed.

Artificial Intelligence

Intelligence is a difficult concept to define, so naturally AI as a field cannot be easily clarified. According to the Oxford dictionary, intelligence is:

"the faculty of knowing and reasoning, the power of thought and understanding"

and artificial is defined as:

"not natural, not real, made by the art of man".

AI can therefore be taken to cover any simulation of a human action which requires some degree of understanding or knowledge to be carried out.

The term "Artificial intelligence" was introduced by John McCarthy in 1965 to refer to the study and development of machine-generated intelligent activity. It is a term which often creates resistance to the concept: implying to some that the aim of AI is to replace humans with computers. However, it is a term which has been apparently accepted and therefore efforts must be made to establish that generally the systems being developed are not intended to replace human expertise – merely support it.

Customarily, computer systems have a huge capacity for storing facts and making them easily available, but the cognitive skills needed to use these facts are largely possessed by humans. Conversely, humans have problems handling large amounts of data, but they can manipulate the data they do possess in very complex ways, enabling them to derive a large amount of additional knowledge.

How do People Think?

Current AI programming is largely intended to simulate or utilise some aspect of human intelligence or knowledge. An accepted advantage of AI systems is that they attempt to represent and manipulate knowledge in a way which can be likened to the techniques which a human would use, while still utilising the traditional computer's capacity for remembering large amounts of data. However, although this is beneficial, we feel that KBS could be used equally effectively in other ways: that is, not to mimic expertise but to support it. These views are discussed more fully elsewhere in this book (see Chapters 9 and 10).

It is very difficult to establish how the mind works: if we knew this for certain then representing and emulating human intelligence would not be a problem! However, knowledge is not knowledge until it is used; idle facts do not constitute knowledge, merely information. It is the way in which facts are used which make them useful. Therefore, the way in which knowledge is controlled and inferenced is as important as the way in which it is represented.

Knowledge has already been discussed in Chapter 2, so it is sufficient to say here that KR must somehow take into account how people utilise their domain knowledge if it is to be effective. However, other factors influence the choice of KR (as will be discussed later in this chapter) and there is inevitably still some confusion and debate about the best form of KR. It is recognised, however, that people think dynamically: they remember new facts while disregarding out-of-date ones. It is sensible that KBS should exhibit similar behaviour and that KR should encompass this.

People can be uncertain about some of the decisions they make. Some researchers feel that uncertainty is an intrinsic part of KR (Cohen, 1985; Gupta and Yamakawa, 1988; Alvey and Greaves, 1990). Others would argue that it is not beneficial to include uncertainty in KBS as people do not think in terms of probabilities or numerical judgements, and as yet there are no other ways of effectively representing uncertainty within KBS (Fox *et al.* 1987; Jones, 1989). The subject of uncertainty has already been covered in Chapter 3.

Different Set of Interests When Developing KBS

When developing KBS which are intended to incorporate or emulate some facet of human behaviour, there is inevitably a change of priorities from those which are apparent when building powerful algorithmic programs. The three major areas of interest in KBS development are (Ringland and Duce, 1988):

- Which development tool?
- Which method of KR?
- Which inference mechanism and control structure?

Essentially these considerations are apparent when developing any computer program: that is, a language must be chosen, some program format must be adopted and somehow the available information must be manipulated. However, in AI, the choice is far more varied and the importance of the choice is fundamental to the success of the project, as it may have a great impact on the overall effectiveness and utility of the developed system, as well as affecting the ease with which the system is constructed.

The three areas of interest identified above are now discussed.

Which Development System?

Basically, this covers whether a language, shell or toolkit is to be used. If a language is chosen, is it to be an algorithmic or symbolic language? Are there any additional programming techniques which are to be incorporated such as genetic algorithms or constraint logic programming? Conversely, if a shell or toolkit is to be used, which one should be adopted?

The range of software which is available has been discussed elsewhere (Allwood *et al.*, 1989; Pham, 1988). However, the range of available software is increasing and changing so rapidly that it is difficult to review it effectively. Therefore, only a brief outline of the different categories of development systems is included here.

The main consideration is to choose a development system which suits not only the needs of the domain, but also your programming skills, resources and time. For example, if the system to be developed is relatively small and the available time is restricted, then a simple shell would probably be the best option. If, however, you are unsure of the specific direction or the eventual size of the KBS (for instance, in a research project) the flexibility afforded by using a language may be beneficial. The hardware which the KBS is to be used on must also be considered. It is no good developing a system which requires a workstation if the only hardware available in practice is a PC.

Languages

The major disadvantage of using a language is that, unless you have had previous experience, it can be time-consuming to learn. However, once the skill of programming in a language has been acquired it can be used for future projects and, also, languages offer maximum flexibility in terms of space, interfaces, inferencing and knowledge base structures.

Symbolic Languages

Symbolic manipulation languages are largely designed for AI applications and are considered to be more suitable than algorithmic languages for work on large scale AI applications (Waterman, 1986; Maher, 1987; Mullarkey, 1987; Van Koppen, 1988).

The two most prominent languages which have been developed are PROLOG and LISP. LISP is based on the manipulation of lists (LISt Programming – hence the name) and PROLOG is based on logic programming (PROgramming in LOGic) which is briefly discussed later in this chapter.

Algorithmic Languages

These are languages such as FORTRAN, BASIC and PASCAL: languages with which most engineers are familiar. They are generally problem oriented and are designed for particular classes of problems (Carroll, 1985), for example, FORTRAN is convenient for numerical manipulation and algebraic calculations.

Despite efforts by many to argue otherwise, algorithmic languages can be used to construct KBS and several systems have been developed in this way. However, for large domains, there are a number of problems associated with these languages: combinatorial explosion, cumbersome programs and more importantly the manipulation of written information can be difficult, as these languages tend to be numerically oriented. Building effective control and search mechanisms can also be problematic.

Overall, developing KBS in standard algorithmic languages is very difficult (Allwood, 1989; Graham, 1989). For more information see Naylor (1983), Hartnell (1984) and Chabris (1988).

C

C is dealt with separately as it is different from both algorithmic and symbolic languages, and it is becoming recognised as a good option for KBS development. The main benefits of C are that it is flexible, offers good graphics and can interact with PROLOG and a number of shells and toolkits. This is because PROLOG is, in fact, built using C, as are a number of other shells.

Fundamentally, C is a procedural language which is similar to PASCAL in the way in which it operates. However, it incorporates additional advantages, such as sophisticated graphics facilities.

C is therefore worth considering if a language is to be used for developing the KBS. An example of a KBS which has been written in C is described in Chapter 9. For more information on C, see Kelly and Pohl (1986), Nelson (1987), Kernighan and Ritchie (1988) and Stroustrup (1991).

Shells and Toolkits

Shells are programming tools which have been specifically designed for developing KBS.

A KBS typically consists of three basic components:

- A knowledge base.
- An inference engine.
- A user interface.

A shell consists of two of these components: a user interface and an inference engine. It also provides a format (or a choice of formats) for the knowledge base. For example, it may allow the knowledge base to be written as rules or as frames or as a combination of these (these KR formats are described later in this chapter). All that is needed to create a complete KBS is to input the information in the correct format.

A toolkit consists of the same components as a shell. When they were first introduced, toolkits were more flexible than shells as they offered a choice of KR and inference mechanism, whereas shells were generally restricted to one option. Recently, however, shells have developed enormously and now offer the same level of flexibility as toolkits. Many toolkits are designed to directly interact with a language (such as PROLOG or C).

Shells and toolkits are relatively straightforward to use and can greatly ease programming, especially when developing a first system. However, they can be restrictive, though this is not as true with modern versions. It can be difficult to choose a shell which suits all of your needs and some can have a very short shelf life: that is, shells are developing very quickly and many are taken off the market owing to lack of sales or the availability of improved versions. This can be a problem for long-term projects.

Overall, shells and toolkits are an ideal way of developing a KBS quickly and efficiently. However, they can be restrictive for research projects or for large domains.

Which Method of KR?

This is the crux of the KR problem. The knowledge in a KBS must be stored in a form which can be readily manipulated by the inference engine.

There are fundamentally three approaches to KR: object-based programming, rule-based programming and logic-based programming. These are described in the following sections. The various members of each category are also included.

When to Choose the Form of Knowledge Representation

There is much disagreement about when to decide on a form of KR. Some believe this decision should be taken at the start of the project before the KE has commenced; others believe that the KR can only be chosen once at least some of the knowledge has been elicited. We have found the latter to be appropriate for design domains. This approach enables the knowledge to be assessed to see which form of KR would most effectively represent it, hence reducing the likelihood of initially choosing an inappropriate KR technique. It also allows the experts to give their opinion on what form of KR seems most suitable. Obviously, if rapid prototyping is used, a form of KR has to be chosen more quickly, and it may be preferable to adopt an approach which is familiar or a shell with which a system can be built quite quickly.

Inevitably, once you have developed a few systems, familiarity is going to play a part in the choice of KR. However, be wary: it is very important to choose the type of KR which suits your domain, and even apparently similar domains may demand different approaches in order to be effective.

Approaches to Knowledge Representation

Now to discuss the feasible approaches to KR. If the considerations involved could be summarised easily, then KR would not be a problem. Unfortunately this is not the case. There are a very large number of considerations when choosing a type of KR for your domain, and this is a problem which has been researched by people in many disciplines including psychologists, cognitive scientists, computer software developers and of course KBS developers. Inevitably, the aspirations and requirements of all these sectors vary and different conclusions are reached by them all. The decisions which they make are inextricably linked to the starting point of their assessment: whether it be psychology, practicality or AI research.

This text is primarily practically based: therefore, it will forgo an in-depth analysis of the psychological issues involved in KR, although they are worth taking time to read as they can help to unravel the web of confusion surrounding the KR issue (Lindsay and Norman, 1972; Newell and Simon, 1972; Johnson-Laird and Watson, 1975) and Chapters 2 and 4 do cover some of the psychological issues which are relevant to this text.

This book will also omit the numerous debates about the type of knowledge being represented: whether it be deep or shallow, procedural or declarative and so on. Knowledge has been discussed in Chapter 2, and any further discussion would be superfluous. Instead, this chapter concentrates on what KR techniques are available. Ultimately, it is up to the developer as to which techniques he or she feels comfortable and confident in using, and which techniques will be most beneficial to the system and to the user.

Object-Based Approaches

As the title suggests, these are approaches to KR which rely on the object as the fundamental component, and manipulation of these objects is how the KR operates. This is in contrast to the rule-based approaches which are described later.

Semantic Networks

Semantic networks were first used in computer programming by Quillian (1968) and since this time a number of semantic networks have been developed.

Semantics is the study of the relationship between words and phrases and their intended meaning. The semantic network is a graphical KR method composed of interlinking nodes, which is intended to represent various meanings, the meaning being dictated by the links (see Fig. 6.1).

The nodes within the network are used to represent objects, concepts or events. Any node can be linked to any number of other nodes, giving rise to a formation of a network of facts. The links between the nodes are called arcs and represent relationships between objects. Commonly used links include "is-a", "has-a" and "is-a-part-of". The arcs can be directed so that the subject–object relationship between the concepts are maintained.

Semantic networks present a form of representation for the knowledge base developer which is relatively easy to develop and understand. New nodes and arcs can be added at any point in the development, although this may prove to be problematic because of the structure of the networks. Prototype structures of semantic networks come very close to that

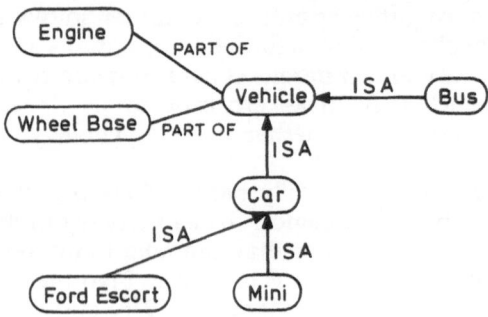

Fig. 6.1 A semantic network

of frames, which are described in the following section and, like frames, the relationships between objects create inheritance hierarchies in the networks. Therefore, in a given network, objects can inherit properties from other objects in that network.

The major restriction of semantic networks is that they are logically inadequate: that is, they cannot express precisely, formally and unambiguously all the interpretations that a human would place on a situation (Woods, 1975, 1983). In addition, semantic networks can only represent predicates with two arguments, so every statement must be in a two-argument form. That is, the two arguments of the predicate are represented by the two nodes and the predicate is represented by the arc which links them. For example:

smith _ _ _ _ _ _ _ _ _ is-a _ _ _ _ _ _ _ _ _ _ > bank manager

where smith and bank manager are the arguments and "is-a" is the predicate.

Another problem is that it is difficult to choose the right set of primitives (that is, the predicates used initially to define the domain) at the outset, and there tends to be an early commitment to a particular range of nodes and links, it being difficult to introduce new primitives at a later stage (Shadbolt, 1989). In addition, there are certain expressive restrictions with semantic networks. For example, qualifications knowledge such as "some" or "all" are difficult to express succinctly. Therefore, despite certain arguments otherwise (Hendrix, 1979) semantic networks are not efficient for large and complex domains. However, they can provide a powerful and flexible base on which more complex hybrid systems can be built.

More information on semantic networks can be found in Brachman (1979), MacRandal (1988), Ringland and Duce (1988) and Shadbolt (1989).

Frames

An alternative form of KR relying on the basis of objects is that which uses frames or schemata. This concept was first introduced by Minsky (1975a, b).

FRAMES & INHERITANCE

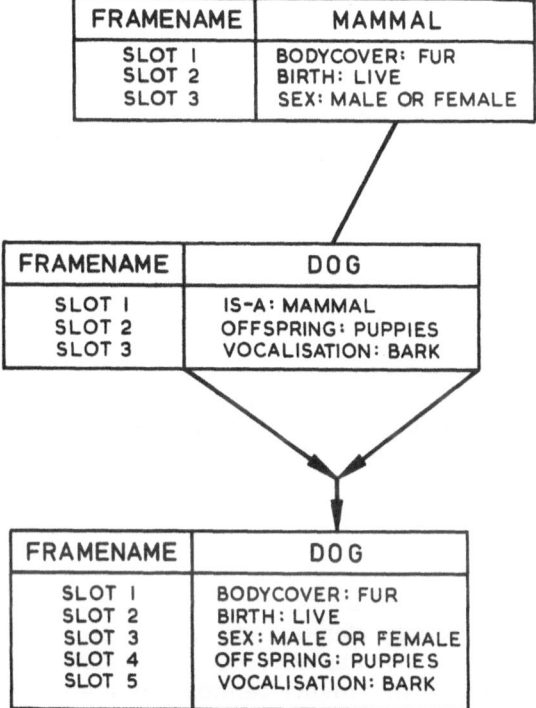

Fig. 6.2 Frames

A frame is a group of attributes which describes a given object. Each attribute is stored in a slot, which may contain default values, rules or procedures for changing the values attached to the attributes. The procedures in a given slot are executed when the information contained in the slot is changed (see Fig. 6.2).

The use of frames assumes that the knowledge can be represented as blocks of data. For example, when dealing with the domain of living things, it is easy to see that one block would be mammals, another would be invertebrates, another would be reptiles and so on. Each block of data is a frame which contains pieces of information, presented as slots. Each slot has the capability of housing specific values about the data or information contained within it. Slots may also contain pointers to other frames or other displays of knowledge.

Frames can be linked together and this linkage gives rise to inheritance: that is, information can be passed down from one frame to another, provided that the frames are linked. For example, if frame 1 containing information about cars in general and frame 2 containing information about the types of cars which are available are linked together, knowledge from frame 1 would be inherited by frame 2. This is a distinct advantage of using frames: that by using this inheritance structure related information can be used repeatedly without the need to develop repetitive blocks of data. Another example is if one frame belongs to mammals and denotes their characteristics (e.g. has fur, breathes, gives birth to live young etc.) then a linked frame on dogs does not have to state the qualities which a dog possesses which are characteristic of its being a mammal, as it can inherit this information from the mammal frame. The dog frame only has to state extra characteristics which are specific to dogs. Similarly, the next frame down on Alsatians does not have to include standard dog characteristics, only those additional features which distinguish that dog as being an Alsatian.

Frames are usually arranged into networks representing the structure of the general concepts within the subject of interest, much like semantic networks: in fact, semantic networks are a form of frame-based system (Ringland and Duce, 1988).

Information in a frame can be represented using both procedural and declarative information. A declarative statement is an assertion that a given piece of information is true. A procedural statement contains instructions to be executed in a given slot. An example of a procedural statement is: "If needed, get owner's address from frame 1".

As their structure suggests, frame systems are useful for domains where expectations about the form and content of the data play an important role in problem solving; for example, interpreted visual scenes or understanding speech. They are particularly effective when used to represent knowledge of certain stereotypical events or concepts (Duce and Ringland, 1988). Default values also play an important part in building frame-based systems. If inheritance cannot play a major role in the construction of the knowledge base, the suitability of a frame-based approach must be questioned.

Frames are ineffective when dealing with a large number of real-world complexities, such as exceptions to the rule, as is often the case in design systems. Effective use cannot be made either of inheritance or default values with this type of domain.

For more information on frames, see Bobrow and Winograd (1977), Minsky (1975a, b) and Ringland (1988). Arguments about the ineffectiveness of frames are put forward by Hayes (1979) and Brachman (1985).

Object Oriented Programming

As the title suggests, object oriented programming (OOP) relies on the principle of using objects, as opposed to rules or logic, as the basis for the KR. OOP is an extension of frame-based reasoning.

Initially, there seemed to be little agreement on the definition of an object oriented programming system (Stefik and Bobrow, 1986; Adeli, 1988a). However, it was generally accepted that in an OOP language such as SMALLTALK, an object is the fundamental computational entity. The only way to perform a computation is to send a message to an object. Objects are organised by class and an object is an instance of a class. The definition of a class determines the structure of its objects. A typical program is arranged into a hierarchy of classes, a class being analogous to an instance of a data structure (i.e. a variable).

The definition of a class may include four types of components:

1. Messages: these are requests for computation and are understood by all instances of the class.
2. Methods: these are attached procedures and are common to all objects of a class.
3. Class variables: the values of these are common to all objects of the classes.
4. Instance variables (analogous to slots in the frame formalism): the values of these are unique to a particular object.

The set of messages that an object understands define its interface (i.e. all the possible ways of referring to an object). For each message, there is an associated method (procedure) that performs the computations necessary to answer the message. Messages (and the methods used to answer them) are inherited from class to class in the following way: if a message is sent to an instance of the class, the message is passed on to the superclass of that class (i.e. the class above it in the hierarchy), and in turn this superclass will send the message to the next superclass and so no. Thus a class inherits all of the messages defined in its superclass or above.

The above constructs offer a number of features, and one of the major advantages of OOP is that it offers a high degree of modularity. Each module is essentially a "black box" which does not have to know the internal structure of the other modules.

OOP is also very efficient for rapid prototyping, as the design of the system does not require connections between the objects to be carefully planned or defined in advance. The only things which have to be specified are the inheritance relationships. Therefore, the addition of new and diverse information is greatly eased.

For more information on OOP, see the SMALLTALK literature, Bobrow and Winograd (1979) or Stefik and Brobow (1986).

Rule-Based Representation

In the KR techniques discussed above, the knowledge is organised around the use of objects as the fundamental computation entity.

A different approach to KR is to express knowledge as rules. Rule-based KR was, until recently, the favoured syntax for the development of KBS. However, it has somewhat fallen from favour because of its apparent restrictions, and preference is now being given to OOP and logic programming. Despite this, it is still a recognised approach to KR and, if programmed effectively, rule-based representation is very efficient for some domains and should always be considered as a possible KR technique.

Rules are generally highly comprehensible to the system developer and whoever is reading the software and, in the first instance, they are essentially easy to program. However, there are considerable problems with issues such as combinatorial explosion and rule ordering which must be considered when larger and more complex domains are being dealt with.

Systems using rule-based architectures are often called production systems, as the rules "produce" a result (Newell and Simon, 1972). The rules are used to represent relationships in terms of condition action pairs:

IF (condition) THEN (action)

for example

IF raining THEN use umbrella

Rules are a classic way of representing human knowledge as the straightforward "IF-THEN" format maps well onto the English language. The rules can contain a number of propositions, for instance:

IF (condition1) AND (condition2) AND (condition3)
THEN (action)

for example

IF sunny AND hot AND humid
THEN use air conditioning

Rules have a readily understandable format, provided that the condition part of the rule does not become too complex. Each rule is only "fired" or considered true if all the propositions contained within the rule are true. If one or more of the propositions is false, then the rule will fail and the entire rule will generally be disregarded; that is, if it is hot and sunny but not humid then the air conditioning will not be used.

The main advantage of production rule systems is that they exhibit useful modularity in that the rules are largely independent of each other and of the rest of the system, each rule encoding a "chunk" of domain knowledge. The overall modularity of these systems and their components helps in construction, debugging and maintenance (Davis and King, 1977; Hayes-Roth *et al.*, 1983; Williams and Bainbridge, 1988).

The explicit representation of rules permits a KBS to readily allow enquiries about rules. Simple chaining methods (as described later) can be used as inferencing procedures which are not unlike those used by humans, and very large rule-based systems can be developed which model expert behaviour in some domains.

A final advantage is the ability of rule-based systems to present records of their own problem-solving processes, that is, "how" and "why". However, it is recognised that in systems built to date this style of explanation is far from adequate, as they are difficult to understand and do not succeed in providing the help which is required, and significantly more powerful explanation facilities are needed (Clancey, 1983; Hughes, 1986).

There are a number of disadvantages with rule-based systems (Williams and Bainbridge, 1988). One of the most commonly recognised problems is that of inefficiency. This has to do with determining the conflict set: that is, the set of rules which are potentially able to fire in a given situation (Shadbolt, 1989). In very large rule bases, conflict sets can become too big to be effective. Various optimisation methods have been proposed (Forgy, 1982) but the problem of applying the chosen conflict resolution rule as effectively as possible still remains. In addition, once optimisation procedures are adopted, the accuracy and reliability of the domain in representing the expertise in question must be considered. Constraint programming is one way around this problem and this is briefly discussed later in this chapter. Alternative solutions are discussed by Davis (1980) and Hayes-Roth (1984).

There are still many advantages to using production or rule-based systems. But to summarise the disadvantages: not all human problem-solving methods are easily represented in the production formalism because of its restricted syntax; control knowledge is often not clear; and management of large knowledge bases can be difficult. The limited syntax and expressive inadequacy have led researchers to consider associated, but more complex representation techniques such as logic programming. However, for simple problems and many domains such as design, rule-based programming is still the most straightforward and in many ways the most beneficial form of KR.

Predicate and Logic Programming

Logic should not be regarded as an alternative form of KR but more as a basis for evaluation, exposition and comparison of the actual notations used in KBS development (Shadbolt, 1989). For example, propositional logic can be seen to form the basis of the rule structure, and predicate logic forms the basis of the PROLOG language.

Predicate logic would appear to be the ideal basis for KR (Black, 1986). It has considerable expressive power for representing the structure of propositions and rules, precise semantics, known formal properties and mechanical inference procedures.

The use of predicate logic for KR has been brought to prominence by the increased popularity of PROLOG. Specifically, PROLOG relies on the Horn clause form of logic which allows it to represent rules and facts in a similar way. The Horn clause structure is:

Conclusion1:- Condition1, Condition2, Condition3.

which can be translated as conclusion1 is reached if condition1 and condition2 and condition3 are true. This illustrates that the ":-" sign represents "if", the commas represent "and" and the full stops represent the end of the clause. Similarly, a semi-colon in PROLOG represents "or".

Logic is critically concerned with the validity of arguments: that is, with methods for determining whether given conclusions can be validly drawn from assumed facts (Frost, 1986).

There have been many disagreements as to the importance of logic in KR (Hayes, 1977; Moore, 1985; McDermott, 1987). Like many debates, the issues have become confused, one of the problems being the lack of a consistent definition of logic itself. In AI, logic is likely to mean one of the following:

- First-order logic (FOL).
- Some development of first-order logic.
- Any formally defined method of representing knowledge and making inferences about it.

The modern basis of logic is first-order logic which is also known as predicate calculus, classical and general logic. Most of the other logics being studied are developments of FOL and inherit at least some of its characteristics.

Logic and its principles are too complex to describe here. For more detail on FOL and associated logics see Frost (1986), Malpas (1988) and Pavelin (1988).

Control and Search Mechanisms

The choice of inference engine can greatly affect the success and efficiency of the system. As discussed earlier in this chapter, in KBS development the inference engine is responsible for deciding how and in what order the data in the knowledge base should be used. It is responsible for the control and execution of the reasoning strategies used in the KBS and is therefore an intrinsic part of KR.

Reasoning strategies used in KBS include forward and backward chaining, which are discussed in detail in this chapter. Breadth first and depth first search are also discussed.

In logic-based languages, such as PROLOG and LISP, the search is controlled by the interpreter or theorem prover, the in-built mechanism which analyses and processes the clauses contained.

Owing to space restrictions, there are many control mechanisms which cannot be covered here. Like the choice of KR, the inferencing procedure should be chosen to suit the domain in question. However, hybrid control structures, that is mixtures between the various techniques, are becoming increasingly recognised, as very few domains of a reasonable size can be effectively searched and controlled using only one mechanism.

Backward Chaining

Backward and forward chaining are control structures used to specify how rules contained in a rule-based system are to be executed. That is, they assign rule, fact or goal order.

Backward chaining, or goal-directed reasoning as it is otherwise known, begins with the conclusion and proves this conclusion by proving the component conditions, giving goal order top priority and working backwards (hence the name).

PROLOG is designed to use backward chaining in conjunction with depth first search, and the overall process is called backtracking. Backward chaining is a good control structure to use when there are more facts than final conclusions, as it begins with the least number of possibilities.

As an example, consider the following knowledge base:

/*FACT1*/   *weather IS sunny*
/*FACT2*/   *distance IS short*

/*RULE1*/	transportation IS bicycle IF
	weather IS sunny AND
	distance IS short
/*RULE2*/	petrol cost IS	zero IF
		transportation IS bicycle
/*RULE3*/	transportation	petrol cost IS zero

Suppose it needs to be established that the journey is free, the question:

transportation cost IS zero?

could be asked.

This goal is not recognised as a fact in the knowledge base, so the rule with a head which matches the clause is used:

transportation cost IS zero IF
petrol cost IS zero

As this is still not a fact which is recognised in the knowledge base, it leads to the sub-goal:

petrol cost IS zero IF
transportation IS bicycle

This is not established as a fact within the knowledge base, so the corresponding rule is found and used:

transportation IS bicycle IF
weather IS sunny AND
distance IS short

This time the two new sub-goals are recognised as facts in the knowledge base and therefore the sub-goal RULE1 succeeds, from this the sub-goal RULE2 will succeed and therefore the original goal is verified as being true.

From this it can be seen that question was used as the starting point, and all the sub-goals which were needed to satisfy this original goal were recognised in turn until the underlying facts were established. These facts were then used to verify the sub-goals and subsequently the original goal: that is, working backwards through the knowledge base.

There are a number of ways to improve the efficiency of backward chaining. For details see Rowe (1988).

Forward Chaining

Often rule-based systems work from just a few facts, and are capable of drawing a large number of conclusions from these facts. In these systems, it is more sensible to use forward chaining which starts with the facts and reasons forwards through to the goals, thus starting with the fewest alternatives.

Using the same example knowledge base as in the previous section, suppose the same question is asked:

transportation cost IS zero?

As before, this question is not recognised as a fact within the knowledge base. In forward chaining, however, the existing facts are established first. That is:

weather IS sunny
distance IS short

are recognised as being true.

Next, the rules associated with these facts are found and the appropriate conclusions drawn. So, from RULE1:

transportation IS bicycle IF
            weather IS sunny AND
            distance IS short

it is recognised that the transportation is a bicycle. The next rule incorporating this information is RULE2:

petrol cost IS zero IF
            transportation IS bicycle

This is still not the information which is required, so RULE3 is used:

transportation cost IS zero IF
            petrol cost IS zero

This is the information which is required, so the original goal is verified as being true.

This is the forward chaining reasoning mechanism, where the facts in the knowledge bas are used to reason towards the goal.

Hybrid Control Structures

"Pure" forward chaining is rarer in applications than backward chaining, but a combination of forward and backward chaining is very commonly used, as this can enable problem solutions to be reached very quickly when dealing with large search spaces. These are known as hybrid control structures.

In most shells and toolkits (and even packaged languages) a choice of backward chaining, forward chaining or some combination of these is feasible. This section has only given a very brief description of chaining techniques. To understand them more fully, it is obviously best to try a sample knowledge base of your own and follow the reasoning path using both mechanisms and a hybrid. For more details see Nilsson (1980), Rowe (1988) and Allwood (1989).

Search Strategies

Inference engines also adopt various search strategies in order to locate pertinent data in knowledge bases. These mechanisms include depth first and breadth first search. As these are two of the simplest search strategies available, they are briefly described here in order to give the reader a feel for the mechanisms involved.

Depth First Search

With depth first search, a start state is chosen, then some successor of this start state (that is, a state that can be reached by a single transition) and then some successor of that state and so on, until a goal state is reached. If a depth first search reaches a state S without successors, or if all the successors of state S have been visited without reaching a goal state, then the state immediately prior to state S is returned to and the search begins again. If the solution space is treated like a tree structure, this mechanism has the effect of searching one branch of the solution tree at a time and only moving back to try another branch when all the possibilities in the branch initially chosen have been exhausted.

Using the example knowledge base:

transportation IS bicycle IF
         weather IS sunny AND
         distance IS short
petrol IS none IF
         transportation IS bicycle
transportation cost IS none IF
         petrol IS none
petrol price IS free IF
         car owner IS someone else
transportation cost IS none IF
         petrol price IS free

the order in which the goals are searched is shown in Fig. 6.3.

If the only provable fact is that the car belongs to someone else, the inference engine would go through six steps before proving the transporta-

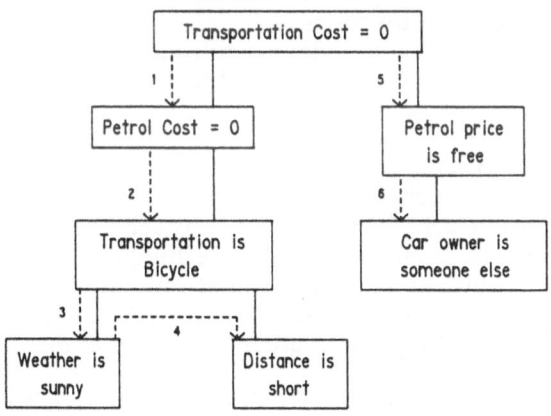

Fig. 6.3 Depth first search

tion cost is none. That is, it goes all the way down the first available branch, before trying the second branch of the tree.

One problem with depth first search is that it works well when searching a tree or lattice structure, but it can get stuck in an infinite loop on other graphs, never finding a goal state.

Breadth First Search

Breadth first search does not have the danger of infinite loops as the solution space is examined one level at a time, all the steps at any given level being examined before going to the next lower level. Therefore, all the immediate successors of the start state are considered before any others, then all their immediate successors are considered and so on, examining the knowledge base one "layer" at a time. Again, consider the above knowledge base: when a breadth first search is used, the order in which the goals are considered is shown in Fig. 6.4.

Unlike depth first search, breadth first search is guaranteed to find a path to a goal state if one exists, but it may take a while as it can be inefficient: always considering extra paths that may not be required. There are ways to eliminate both the looping difficulties of depth first search and the inefficiency of breadth first search, but they are too complex to describe here. For more information see Rowe (1988).

Depth first and breadth first search, often occur in disguise in AI applications. The backward and forward chaining algorithms described earlier are really types of depth first search and are therefore depth first control structures. These control structures are common in computer applications because of their easy-to-implement data structure.

Other Search Techniques

A major concern of AI research is to find better search strategies. Many of those found are related to depth first and breadth first search strategies. Two of the most commonly used variants are search using heuristics and search using evaluation functions.

Heuristic search techniques employ rules of thumb to help solve problems and can be applied to both breadth first and depth first searches. These techniques use some intelligent scheme to reduce the size of the search space; that is, they "prune" the solution tree by eliminating branches. With these search techniques, often only part of the solution space is searched, so there is no guarantee that the best solution for a given problem is reached: only that a satisfactory solution will be identified. They are essentially optimisation techniques, as they limit the area of the domain which is searched so that only an optimum solution for the domain area searched is reached. Heuristic searches include means analysis and hill climbing. For more information on these see Martin and Oxman (1988).

More recent developments for reducing the search space include genetic algorithms and constraint logic programming. These are briefly

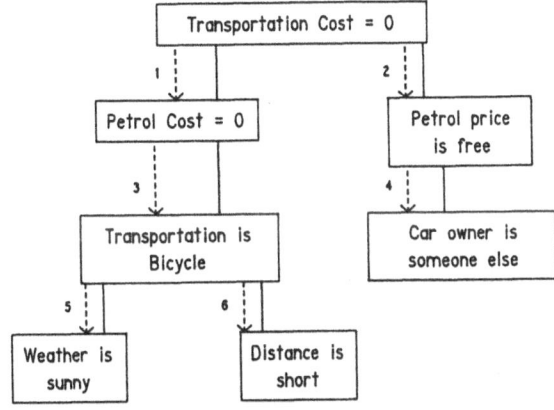

Fig. 6.4 Breadth first search

described here, but obviously many of the important issues attached to them are omitted.

Genetic Algorithms

Optimisation by some form of random search is often preferable to standard optimisation techniques (such as hill climbing methods) when the domain in question contains multi-modal and non-linear functions. In addition, these techniques are only able to optimise one part of a complex system of interdependent processes. Undirected random searching is very inefficient for large domains as it will, on average, only cover half of the area in an attempt to locate an optimum solution.

Genetic Algorithms (GAs) are a way of reducing the search space by carrying out a directed random search in order to reach an optimum solution. They emulate the processes of genetic evolution found in nature, by using a random initial population of possible solutions and progressively improving each "generation" of solutions by artificial "breeding". As the GA evolves, it acts as a store of past generations, as it contains the genetic elements of "fit" solutions that have survived a number of generations. The fittest of the "final" generation should be an optimal (but not necessarily "the" optimum) solution.

The main advantage of GAs is that they are extremely robust and do not require information about the nature of the objective task which they are optimising. GAs can be slow, but they are ideally suited to parallel processing, and they can optimise in situations where there is no other viable option.

Although GAs have been around for at least 15 years, the field is still in its infancy and much research is still being carried out in this area. Many potential areas of application have been suggested, but they have not yet been totally accepted. The current applications can be broken down into three areas: experimental, practical and in the development of "classifier" systems. These are knowledge induction systems with an adaptive

(generic) component. Examples of practical applications include scheduling and layout problems and image and signal processing (Etter and Dayton, 1983; Fitzpatrick *et al.*, 1984; Baffes and Wang, 1988; Sponsler, 1989).

GAs have also been applied to rule-based expert systems for learning.
These cannot be ventured into here, but the following references provide
the necessary information: Rada *et al.*, 1984; Schaffer and Grefenstette,
1985; Holland, 1986; Soucek and Soucek, 1988.

Constraint Logic Programming

This is basically as it sounds, i.e. "constrained search". Constraint logic
programming combines the declarative aspects of logic programming with
the efficiency of constraints (Jaffer and Lassez, 1987). The logic is used to
specify a set of possibilities which are explored using a very simple built-
in search method (e.g. PROLOG, which uses an in-built backtracking
mechanism as mentioned earlier). The constraints are used to minimise
the size of the search. These constraints can be used to limit the search
space in a number of ways. For example, for:

- Consistency checking (i.e. making sure that routes are not followed
  which exhibit inconsistencies).
- Lookahead procedures (which check that the next route is worth pur-
  suing).
- Delay mechanisms (i.e. a natural way of coroutining the execution of
  several logic programs by specifying when a predicate can be selected
  for execution).
- Local value propagation (each constraint states a relationship among
  variables, so that an appropriate value can be computed once the remaining variables in that expression have been instantiated).

By using constraints in this way, the search can be made more efficient
while ensuring that all the viable possibilities have been considered.
Therefore, it is not an optimisation method: it does in fact look at the entire knowledge base, so that you can be sure that a best solution is
obtained.

The main application areas for constraint logic programming are inevitably areas where constraints play and important part in the domain, such
as planning and design domains (Simonis and Dincbas, 1987; Dincbas *et
al.*, 1988).

Examples of constraint logic programming languages are CHIP (Constraint Handling In Prolog) (Dincbas *et al.*, 1988); PROLOG III (Colmerauer, 1987); and CLP(R) (Jaffer and Lassez, 1987).

Rule Selection

As well as the above techniques, the search pattern can be controlled by
dictating the selection of rules. There are two basic approaches to this: by
using meta-rules or decision lattices.

Meta-Rules

Meta-rules are rules about rules (Davis, 1980, 1982). They are a special class of rules which represent control knowledge and prescribe the manner in which rules should be applied. Meta-rules are apparently used in human reasoning, and their importance in effective KR is recognised. However, actually eliciting these rules is very difficult.

Meta-rules can be used to break down a knowledge base into segments, where each segment applies to a certain class of situations. They can also be used to provide strategies for the selection of useful paths of reasoning. Meta-rules are also be used to give priority to relevant parts of the knowledge base (referred to as focusing). As well as establishing priorities they can enable tangible lines of reasoning to be constructed. Another use of meta-rules is for re-ordering rules and facts within the knowledge base when appropriate. They are therefore particularly useful when dealing with large, single-subject knowledge bases. In fact, when dealing with a KBS of considerable size, the use of meta-rules is virtually inevitable in order to maintain the efficiency of the system.

A simple example of meta-rules are rules which decide how to load partitions (that is, sections or modules of a system), but they do have many other uses. Some examples of simple meta-rules are:

- Prefer the rule which handles the most serious issue.
- Prefer the rule which was obtained from the most reliable source.
- Prefer the rule which is quickest to execute.
- Prefer the rule which has been used successfully most times.
- Prefer the rule with the most things in common with the last rule applied.

Meta-rules can also be used for default values: providing overriding answers when the given answer is highly unlikely or improbable.

Decision Lattices

Decision lattices are a much lower-level form of rule control. They are essentially an approach to system building as opposed to a control structure, but it is an approach which directly affects the control of the system, which is why they are included here. They undertake a very restricted but very efficient style of reasoning. The idea is always to specify where to go next in the system based on previous question answers. Many people do not consider them to be true AI, but they are a form of control. They can be useful for simple KBS as they are easy to implement, and they can be used to partition the KBS so the absolute minimum number of questions are asked.

However, decision lattices have major disadvantages. As a low-level control structure they cannot reason properly, because they do not permit variables of backtracking; they are difficult to modify and they cannot re-use answers (Rowe, 1988).

Decision lattice are therefore rarely used now as they have largely been abandoned, and more sophisticated methods are preferred.

Conclusions

The main conclusion is that there is a wide range of KR techniques and control mechanisms available: both must be chosen to suit the size and nature of the domain being dealt with. No clear "best" method of KR has yet emerged and the issue of which KR to use is still confusing. This chapter has identified some of the important considerations and the alternatives which are available.

From the discussion in this chapter it can be seen that KR can be broadly divided into three areas: which building environment, which KR format and which inference and control mechanisms to use.

The building environment is fundamentally a choice between some form of KBS building tool or a language.

The KR format relies on a choice between representing the knowledge as rules, in logic or as objects. All have their relative advantages for varying domains.

The control mechanism is established as being equally important as the form of KR chosen. This will again be inextricably linked to the domain being used and the choice of KR. Many of these techniques are inefficient when used on their own, so some combination of approaches is recommended for large domain structures. Similarly, there are certain ways of restricting the search space and these should be considered for increasing efficiency when searching large domains, as is usually the case in design-based systems.

Chapter 7

Evaluation of Practical Systems

Objectives

As the aim of this book is to discuss the development of practical KBS, it is only sensible to include a chapter on evaluation. After all, without effective evaluation it is highly improbable that the KBS developed will be of use in the real world.

Therefore, this chapter aims to:

- Discuss practical KBS evaluation: its relevance, the difficulties with it and why it is important.
- Discuss the ways in which systems have been evaluated in the past and to suggest ways of improving this process.
- Suggest new approaches to practical system evaluation.

Introduction

Evaluation is inevitably an important part of the development of a KBS which is intended to be implemented in a practical environment as opposed to being a research tool. Evaluation of any software is a difficult and tedious process, but unfortunately KBS evaluation is an even more complicated task, owing to the nature of the information contained (Partridge, 1988; Allwood, 1989; Bramer, 1989). Currently, KBS generally aim to encapsulate or assist some aspect of human expertise by incorporating expert knowledge on a limited domain. Consequently, the knowledge bases tend to be subjective and context dependent, particularly in design domains. In addition, the knowledge bases are often very large, containing a very high number of feasible solutions or diagnoses. These factors make the system especially difficult to evaluate.

Despite the importance of software evaluation, little work has been carried out on evaluating KBS. This is partly because a large number of systems have not been developed beyond the state of being research prototypes. As it is never intended to use these systems in practice, the importance of evaluating them is reduced. However, according to the available literature and reports on existing systems, even the systems which are developed with practical implementation in mind are rarely put through a rigorous evaluation process.

If practically oriented systems are to be of use in the real world, the importance of detailed evaluation must be recognised. Systems which are to be adopted within a working environment should not only be theoretically correct, they should also successfully and reliably carry out the tasks required by the intended user. The only way to achieve these goals is to test the effectiveness of the system by evaluating it in a practical environment.

Recommendations as to the best approaches for verification and validation of KBS have been put forward by a number of people (Welbank, 1983; Partridge, 1988; Furse, 1989), but these recommendations are not generally based on practical experience or evaluation of working systems. This chapter aims to give a more practical guide to KBS evaluation, the ideas put forward being based on our experience in this area.

Detailed KBS evaluation is not an easy task. It demands a great deal of time and effort from both the system developers and the evaluators, careful planning and hard work being required to ensure the evaluation is productive.

In this chapter we suggest some viable approaches to KBS evaluation. It is recognised that these are by no means the only techniques which can be used, but they are methods which we have used and found to be effective. The approaches which we describe also give a good indication of the considerations involved in practical system evaluation.

Evaluation, Verification and Validation

Firstly, what is meant by the term evaluation? The appraisal of any system can be considered to consist of three components: evaluation, verification and validation, as shown in Fig. 7.1.

Evaluation

Evaluation is the term given to the appraisal of the overall value of the system (O'Keefe *et al.*, 1987). It incorporates the verification and validation components, as well as an analysis of such aspects as overall performance, feasibility of expansion and user acceptability. The evaluation process also involves an appraisal of the system as a whole, assessing such things as the user interface, help facilities and speed of performance.

Validation

Validation is a very important part of the evaluation process which is often confused with verification. Validation substantiates that the KBS performs with an acceptable level of accuracy, that is, it ensures that the system gives a correct answer.

SYSTEM ASSESSMENT

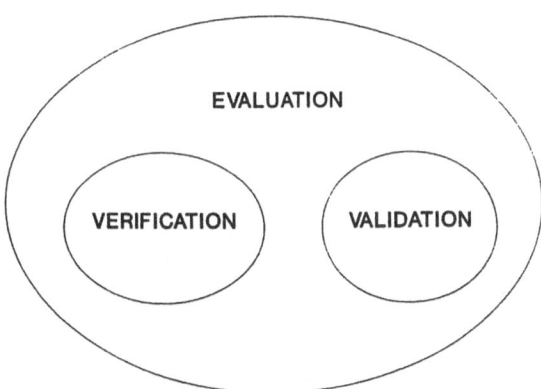

Fig. 7.1 Evaluation, validation and verification

Many of the KBS developed to date have emphasised the program's performance at its decision-making task in evaluation studies (Gaschnig, 1979; Gaschnig *et al.*, 1983; Yu *et al.*, 1979a, b). Since reliably accurate advice is essential for any KBS, this emphasis is logical as highly efficient implementations of invalid systems are useless in practice (O'Keefe *et al.*, 1987).

Verification

Verification can be taken to be the process of ensuring that the system reaches the conclusions it makes in a correct way: that is, ensuring that the system correctly implements its rules and specifications.

This area is also an important part of evaluation but it has received less attention than validation. If a KBS is to be successful, its reasoning processes are inevitably important as it is equally important to verify the reasoning procedures adopted by a system as it is to validate its content.

Why Bother with Evaluation?

It has previously been argued by some investigators that KBS evaluation is not worthwhile because, although the research into KBS development is now past its infancy, it is still actively developing. Arguments have been put forward that as KBS domains are highly subjective and therefore can never be fully corrected or verified, evaluation on a large scale is pointless (De Millo *et al.*, 1979; Smith and Baker, 1983).

Recently, however, this viewpoint is changing. If KBS are to have a real future in commercial environments, then evaluation is essential. Without it, it is unlikely, if not impossible, that their development will be taken seriously because people will not accept a tool whose reliability has not been tested.

An additional argument for formal evaluation is that a system will be continually evaluated regardless of whether this evaluation is planned. Inevitably, when a KBS is put into a working environment it will be evaluated in some form by the people who use it. If this evaluation is carried out informally and the only result is rejection of the system because it is poorly directed or there is a lack of contact with the system developers, then this is surely detrimental. It seems preferable to carry out a formal method of assessment so that some benefit can be gained from this evaluation, enabling the system to be improved and consequently increasing its acceptability.

An added incentive is that formal evaluation instigates feedback from the users, helping to ensure that the program is developed in a way which is suited to its intended environment. Without evaluation, mistakes in the development would not be noticed and the same errors could be made in new projects. Evaluation can help by identifying such things as inappropriate knowledge representation approaches and search mechanisms, as well as clarifying the human knowledge encapsulated within the system.

It is sensible to anticipate that extended evaluation can help the experts as well as the developers (Gaschnig *et al.*, 1983). As experts evaluate the content of a KBS, they are logically forced to evaluate their own expertise. This can help them to structure and gain a better understanding of their knowledge.

A long-term consideration is the inevitability that as KBS become more widely accepted they will be developed in numerous practical areas. Research into evaluation is important so that more effective methods can be developed before practical KBS implementation and consequent evaluation becomes widespread.

Why is Evaluation Important for Design KBS?

Evaluation of any KBS is important, but in a highly practical field such as design, it is particularly so because designers have to make artefacts which both work and are reliable They can therefore ill afford to risk using a system whose correctness and reliability are not proven.

When approaching an design project, designers generally find it necessary to have some proof of validity of the technique or tool being used. For example, it would not be sensible to expect an engineer to design a bridge using newly developed components which had not already undergone thorough testing. It is logical to anticipate that designers will expect the same standards in the development of a tool such as a KBS. A system which has not undergone a fairly rigorous evaluation procedure would not be readily accepted within a design environment.

Realistically, in the future, as design is a dynamic field which constantly changes somewhat with the fashions, successful design systems will need to be expanded. This expansion is neither possible nor sensible unless the original system has been thoroughly evaluated.

Difficulties with Evaluation

As mentioned earlier, evaluating any significant computer system presents difficulties, but these difficulties are accentuated in the case of KBS (Gaschnig *et al.*, 1983; Bramer, 1989).

Bearing in mind the large number of possible routes available through a relatively small knowledge base, it is unrealistic to believe that a KBS will ever be fully tested. This is not, however, a reason to ignore evaluation altogether. Even a small amount of evaluation can help to ensure that correct answers are given at least some of the time, and the longer and more intense the evaluation period, the more accurate the resultant system is going to become. Surely, this is preferable to no evaluation at all! Similarly, once the system has undergone evaluation and the developers are aware of the errors and gaps in the system, they are going to be more receptive to future criticism of the system.

Evaluation is also important as it is highly unlikely that the knowledge base encapsulated will ever contain all the experts' knowledge on the target domain. Evaluating the system can help to ensure that more of the "missing" knowledge is identified.

An added difficulty with evaluation is that for many KBS domains the "correct" solution is not known (for instance, in medical diagnosis) or is not "knowable" (e.g. in deciding a company's long-term financial strategy). Engineering is no exception, particularly in design, where invariably there is more than one feasible solution. For example, when recently visiting a company who were interested in viewing the bridge design system, we found that they preferred to use bank seats which were positioned half way up the slope to using bank seats at the top, which was contrary to the opinions of the experts which we had previously encountered. Consequently, they rejected a large number of the solutions which were suggested by the system, as this was not a preference which had been included in the knowledge base. This is another factor which makes it very difficult to assess whether a system's conclusions are right or wrong. The most that can be concluded is that if the system's solutions differ from the general consensus of the experts then there is probably an error in the knowledge base. However, the differences in opinion between the experts and indeed the expert's own fallibility must also be taken into account. Bearing all these things in mind, a demand for absolute correctness is therefore impossible, if not pointless in the majority of situations. One of the major difficulties with KBS evaluation is gauging the level or performance which it is reasonable to expect the system to reach.

There is always the possibility that companies will refuse to accept systems which are not 100% guaranteed. This will only mean that these companies will never be able to implement KBS, which would be an irrational decision as these companies would not refuse to employ human experts on the grounds of their fallibility or the difficulty involved in testing their performance (Bramer, 1989).

A further complication is that as confidence in KBS implementation in industry increases, the system will inevitably be applied to the solution of problems which occur very rarely – where human expertise is fragmentary and widely dispersed. Inevitably, these KBS will be very difficult to validate, but the encapsulation of such rare expertise is still beneficial.

Even in situations where absolute solutions can be determined, errors will still occur. It is important to realise that not all errors are equally significant. A system may well be forgiven minor occasional errors, if it generally performs to the same standard as a human expert. However, one major error in an area such as medical diagnosis is likely to be sufficient to justify rejecting the system altogether, as may well be the case with a human expert. However, it is also realised that people are far more willing to forgive errors in humans than they are computers: maybe this is because we recognise our own fallibility and are therefore more prepared to accept it in others, yet we find it hard to accept that computers are equally fallible in certain areas.

How these Difficulties can be Overcome

There is no easy solution to these problems. The number of difficulties apparent in evaluating KBS would seem to imply that consideration of correctness needs to be replaced by a consideration of acceptability. Therefore, some changes in both approaches and overall attitude to evaluation may be needed.

The new assessment techniques could well be based on the techniques currently used to judge human expertise. For instance, it is normal for newly graduated engineers to make mistakes. It is unlikely that they will be trusted with a project of any importance within the first few years of their career. However, after a period of working successfully (but not necessarily without error) under supervision, the confidence in these engineers will have increased dramatically, enabling graduates to rise to a higher level, with a greater degree of trust and responsibility.

Therefore, one approach to try and overcome the difficulties with evaluating KBS would be to classify the standard the KBS has reached (Bramer, 1989). The primary problem with this approach is that human experts are seldom evaluated objectively (Gaschnig et al., 1983). Some would argue that the same standards cannot be applied to machines and people. Human experts must usually pass some test or standard to qualify as experts, but experts are not selected according to a numerical rating system. There has been much research on human testing, but as yet little of this research is directly applicable to the evaluation of KBS.

New techniques for KBS evaluation are required. These techniques can only be developed from practical experience. If new techniques can be successfully formulated, it can be anticipated that they will also be applicable to the evaluation of human expertise.

Evaluation of Other Systems

In the past, KBS evaluation has not been given much attention (Gaschnig *et al.*, 1983; O'Keefe *et al.*, 1987; Bramer, 1989). Typically, KBS performance has been evaluated by purely running test cases through the proposed system and comparing results (that is, the classification, final certainty factors and advice given) against known results and expert opinion. Then a percentage of the system's success rate is calculated and subjective judgement is used to analyse the mistakes and contradictions. Examples of this approach range from MYCIN's early validation (Yu *et al.*, 1979a) to the reported validation of EMERGE – a chest-pain diagnosis system (Hudson *et al.*, 1984). Although the use of case studies is a very important part of the evaluation of KBS, it is not sufficient on its own. Case studies need to be supported by other evaluation techniques which help to evaluate the accuracy of the reasoning processes adopted, the suitability of the system to its intended user and the practical applicability of the system.

The simple approach of using purely test cases or case studies presents a number of problems. The success rate percentage is highly dependent on the choice and number of test cases used. When the KBS is compared to the experts on whose knowledge it is based, this type of validation has been found to be of questionable value, as was noticed in the validation of Prospector (Gaschnig, 1979). In all the above mentioned cases, none of the other aspects of evaluation, such as an appraisal of user interaction or an assessment of the utility of the system, were carried out.

Many systems begin as research prototypes and consequently validation has often been conducted to measure qualitatively system performance. This was the case with Internist-1 (Miller *et al.*, 1982). The EDP-Xpert (Hansen and Messier, 1985) and Casnet (Kulikowski and Weiss, 1982) were evaluated to assess their overall value in a particular domain as opposed to being validated or verified.

More complex techniques, such as statistical analysis and formal validation methods, are rarely used for KBS evaluation. Two examples of where this approach has been used are in the later evaluation of MYCIN (Yu *et al.*, 1979b) and Onocin, a chemotherapy adviser (Hickam *et al.*, 1985). The VAX configuration system also underwent some formal validation (Bachart and McDermott, 1984). Using formal methods is inappropriate for the evaluation of KBS, as these approaches are not relevant to the evaluation of human expertise, and, as discussed earlier, it is important to assess KBS on a similar level to human experts (O'Keefe *et al.*, 1987).

Considerations in Evaluation

From the above definition of evaluation and also from the deficiencies which have been noticed in the evaluation which has been carried out in other systems, it is obvious that evaluation must incorporate many factors if it is to be successful. As well as validating the knowledge and verifying the reasoning employed by the system, a number of other considerations must be taken into account if a KBS is to be effectively evaluated. The following sections describe these considerations and identify how they can be dealt with in practice.

Objective Standards

Most new techniques need some kind of gold standard: that is, a generally accepted correct example with which the results of the new methodology can be compared. An example of such a "gold standard" is the Turing test. This is a test invented by Alan Turing in 1950 and has been used by many KBS developers as the test which determines whether or not a computer system is truly "intelligent": unfortunately as yet no system has truly passed the test, although claims exist that some systems have. ELIZA (a psychological advisory system) has come very close to passing this test. For more details see Weizenbaum (1965) and Partridge (1988). As the Turing test is so demanding, and it is now generally accepted that "intelligent" computers are not necessarily the ultimate aim of many KBS, a more realistic standard has been adopted for testing the average computer. This standard is taken to be if it can be demonstrated that the system's advice is comparable to that which would be given by human experts.

Bias

Obviously, as discussed in Chapters 2 and 5, bias must taken into account, as the system may be developed and evaluated by biased experts. Bias is going to be present in nearly every expert; but it can be tempered by using more than one expert and by being aware of its presence.

Bias can also be introduced by the developers, for example, by choosing test cases which guarantee good performance. To avoid this effect, allow the reviewers more control over the system evaluation, such as letting them choose the test cases, to ensure a good cross-section are used.

When judging KBS performance, an expert biased against computers may unfairly assess the KBS. The danger of ignoring this consideration was shown in the evaluation of MYCIN (Yu et al., 1979a), where many of the reviewer's comments reflected their views on the "proper" roles of computers in medicine. This bias can be somewhat eliminated by carefully choosing the experts.

We have encountered a similar problem when demonstrating the bridge design system to a new company. This company were under the impression that the bridge system would cover all parts of the design process and would therefore singly produce a completed and absolutely correct design. This is not the case: the design system only deals with road bridges crossing a road and it only deals with conceptual design. It does not aim to give a complete design and it is recognised that although it provides a correct answer, this answer is not the only feasible solution. However, as the company we were dealing with had a false impression of the system's capabilities, they were very disappointed in the system's performance. They therefore became very negative in the system's review, trying to catch the system out with unusual cases and cases which fell outside the system's domain. They refused to accept any of the answers which were produced, even though some were viable solutions, preferring to believe that the system could not deal with any aspect of the design process successfully. This is an example of reviewers whose bias against a KBS influenced the evaluation of it.

Overall, bias is an unavoidable problem. The best way to counteract it is to be aware of its presence and to try and take its effects into account.

User Orientation

It is obviously important that potential users are involved in the evaluation process. Through evaluation, the users can test a system's competence in its domain of expertise and also determine whether the system produces results which are meaningful to them as well as to the experts used to develop the system.

If the system is to be accepted in practice, it is important that the users' interaction with the system is considered, that is, the system's facilities for assisting them in using it, methods for input of knowledge and output of results, and methods for transmitting information and help to the user. The user can decide which capabilities are useful, what others are required and what can be ignored. The importance of user involvement is emphasised throughout this book, and this topic is discussed further later in this chapter.

Efficiency

The impact of the proposed system on the users' environment must also be evaluated if the evaluation is to be complete. For example, a system that requires an excessive amount of time from the users may fail to be accepted in its working environment, even though its content is flawless. The time a system takes to run relative to the results it provides must therefore be taken into account, as inefficient working practices will not be tolerated.

Interaction of Knowledge

Preserving good performance while correcting the content of a KBS is important in progressive evaluation. Problems can be encountered when evaluation has revealed system deficiencies and new knowledge has to be added in order to correct them. In complex systems, the interaction of new knowledge with old can be unexpected and lead to detrimental effects throughout the system. An awareness of the potential problem is vital and the only way to counteract it is to test thoroughly the system once any changes have been made. This task can be both time-consuming and frustrating. Automated tools are gradually being developed for testing such aspects of development which will make the process much easier.

Realistic Time Demand on the Evaluators

Although this may seem a mundane issue, the amount of time which is demanded of the reviewers is nevertheless important. An evaluation period which is considered too long can be detrimental, as the reviewers can lose interest, perhaps resulting in unproductive evaluation. At the very least it can lead to unacceptable delays in the assessment.

For some domains, one way of overcoming the problem of excessive time demands is to choose two or three pertinent issues to be assessed by the experts so the time required is reduced. However, this approach would not be acceptable in a field such as design, where an overall assessment of the entire system is essential to ensure that the design produced is adequate.

Just to emphasise the importance of this consideration, the problems which can occur with unrealistic time demands were shown in the evaluation of MYCIN, where over a year's delay was faced in obtaining the evaluation information, because the time demanded from the evaluators was more than they were able to commit. The evaluators involved were busy people: if the evaluation had demanded less time from them they would have been able to complete the evaluation more quickly. As it stood, the only way they could complete the task was in bits and pieces when the time became available and as their workload allowed.

An Overall Summary of the Basic Approaches to Evaluation

Now that the approaches to evaluation which have been adopted by other people have been reviewed and the main considerations involved with evaluation identified, this section will summarise the fundamental evaluation techniques which have been mentioned. The following sections will then describe in more detail the type of evaluators and the approaches to evaluation which we have used and found to be successful.

Using Case Studies

The importance of using case studies for KBS development is well recognised (O'Keefe et al., 1987; Bramer, 1989).

In an ideal world, a large number of test cases would always be available for evaluation purposes, these test cases covering solutions given by a number of experts on a complete range of problems. Obviously this does not happen in reality, where usually only a small number of test cases are available, covering a limited range and representing the thoughts of only a few experts. In extreme situations, there may be no test cases available at all.

The choice of case studies used can bias the success of the evaluation. Any cases which have been used in the system development should be discarded from the evaluation, as it can be assumed that the system has been built to deal with them successfully.

For a fair cross-sectional evaluation, test cases should be selected at random; preferably not by the system developers. Obviously, knowledge of the number of test cases used affects the users' confidence in the system, but the important criterion is not the number but the range of cases selected and how well these test cases cover the input domain of the system.

The input domain is the population of the permissible input as opposed to the application domain in which the system operates (Hetzel, 1973). The larger the input domain, the more difficult the evaluation becomes.

The randomness of the test cases can be maintained by allowing the reviewers to choose them. To ensure that the opinions of a number of experts are catered for, where possible test cases from a variety of sources should be used.

Using Experts

Any KBS should obviously initially be evaluated by the experts who were used to develop it. An independent review, using experts who are new to the system, should also be included in the evaluation as this helps to ensure that the knowledge base is more objective and is reviewed in detail. New reviewers have a greater interest in evaluation and are therefore more likely to offer a greater contribution. The reasons for this are explained in the following section.

Using a Broad Range of Interviewing Techniques and Reviewers

In addition to using just experts or purely case studies, there is another option of using a broad range of reviewers, and approaches. Naturally, this is the type of evaluation process which we prefer as it enables more aspects of the evaluation to be covered (i.e. the considerations which were stated in the above sections).

The types of reviewer who should be used include experts, experts who are new to the system, potential users, and other designers or people who may be interested in the system's development. As well as using case studies, the range of review techniques can also include user trials of individual parts of the system, so that these parts can be evaluated before they are incorporated in the overall system.

Reviewing parts of the system can be especially effective when trying to improve certain facets of the system. For instance, when developing a new user interface for the bridge system, one screen of a number of different types of interface style were developed, as opposed to developing complete systems using the new interfaces. This meant that these interfaces could be easily tested using trials, and then the most popular (or a combination of the most popular) interface actually be incorporated into the system. This is a much more efficient and less time-consuming way of evaluating certain aspects of the system. A similar approach could be used for evaluating the help system which is included in a KBS. However, it must be recognised that this approach is only useful for evaluating parts of the system which do not actually affect the solutions or advice which the system provides. The system as a whole must also be evaluated if a true idea of the system's correctness, acceptability and practicability are to be acquired.

The Reviewers

As has already been stated, the experts used to develop the system should be the first people to review it, for a number of reasons. Firstly, because they are more likely to spot glaring errors as they are the people most familiar with the knowledge base. Secondly, it gives them the opportunity to give their opinion on the system: its overall appearance, behaviour and representation of their reasoning processes; so if they feel that the system does not fairly or accurately depict their expertise, it can be changed to incorporate their suggestions before any external reviewers see it. This is important as it protects the integrity of the original experts as well as emphasising their importance in the project.

Once the original experts have seen and approved the system, it can be sent out to a number of external reviewers, that is, expert engineers who have not previously been involved in the system development. These reviewers are independent, helping to create a more objective knowledge base, and their involvement helps to ensure that any gaps, bias or mistakes in the knowledge base are eliminated. It is important that reviewers who are new to the system should be involved in the evaluation as they are more likely to be enthusiastic about the evaluation, feeling that they have something concrete to contribute. Unfortunately, often the experts who were used to develop the knowledge base lose interest in the system once they have confirmed that the system gives a fair representation of their knowledge. Their attitude towards the evaluation then tends to become apathetic as they feel their task is complete, and this leads to ineffective evaluation.

Reviewers who are new to the system tend to be more interested because they have not seen the system before and because they want to find something which they can change about the system so that they can prove their worth. Also, they are fresh to the project and are therefore not bored with the system development, as is often the case with the original experts. Once the reviewers realise that their comments and criticisms are being taken notice of, generally their enthusiasm for the review process increases. Once they see that the system is being changed to suit their needs and ideas, they think harder about their comments and are more willing to contribute to the evaluation process, and in our experience many reviewers actually look forward to seeing the "new" versions of the system.

In order to make the evaluation more general, a broad spectrum of reviewers should be involved. This means not just potential users and experts, but also people who have a range of qualifications in between, and who may be interested in the system's development. For the systems which we have evaluated, engineers from all backgrounds were used: ranging from engineers with no previous design experience to fully qualified "expert" designers.

For example, in the bridge design project, the potential users were identified as being young engineers with no previous design experience. During the evaluation, this type of engineer was obviously involved, as were expert engineers. However, in addition, a number of junior engineers were included. These engineers had experience in bridge design but could not be considered to be fully experienced in their field. This meant they could evaluate the content as well as still being able to appreciate the needs and opinions of the users. By using this "interim" type of reviewer, the applicability of the system could be extended, as these reviewers were able to identify areas suitable for expansion and aspects which could be added to the system to make it more applicable to their areas of work.

Where possible, it is advantageous to include more than one company in the evaluation of any system. This enables a broader spectrum of opinion to be considered as well as helping to ensure that the system is not developed to suit only one design office. The suitability of the system to a number of practical environments can also be analysed, ranging from small offices to large international companies, and the alterations which would be required to adapt the system to suit these environments recognised. Of course, if the system is intended to be used in only one office, then this does not apply.

Just as an additional note, if different types of reviewer are to be used, be careful, because inexperienced users can feel intimidated by the presence of more experienced engineers, making them less likely to contribute to the review process. One effective way of overcoming this is to arrange to meet reviewers in groups of similar status. Then, they will be more likely to contribute to the discussion.

The Evaluation Process

The following sections describe the evaluation process as we see it: inevitably this description is based on our experience in system evaluation, so personal opinion influences the discussion.

When to Evaluate

Little agreement currently exists as to when evaluation should take place. Typically, a KBS must exhibit an acceptable level of performance at some early stage of development, but this level of acceptance will vary according to the intentions of the system. For instance, the expectations of a prototype differ from those of a system intended for commercial use.

It is our opinion that KBS evaluation should be continuous and, once the preliminary verification and validation have been carried out, a more structured evaluation can be used. Therefore, evaluation occurs throughout the system development, becoming more structured as it progresses, in much the same way as the KE process.

A major influence on the type of evaluation methodology which should be adopted is whether rapid prototyping is used for the system development. As discussed in Chapter 5, whether to use rapid prototyping or an alternative development strategy is an important decision, which is largely dependent on the type of project being tackled.

If rapid prototyping is not used, detailed evaluation of the system can begin as soon as the first prototype has been built. Inevitably, if rapid prototyping is used then a form of evaluation will start virtually as soon as the project begins.

Now that the considerations in evaluation have been identified, the next stage is to suggest a strategy for the evaluation process. As has already been stated, this strategy is based on our experience in this area; and approaches are suggested which we have found to be successful. Inevitably, many other approaches to evaluation could be used, but we have no way of measuring their appropriateness or applicability.

Preliminary Evaluation

The preliminary stage of the evaluation aims mainly to check that the information contained within the system is correct and complete, ensuring that any obvious pieces of the knowledge base which have been missed are elicited. Obviously, this stage also aims to identify any bugs in the system.

This stage of the evaluation also hopes to identify any aspects of the system which are confusing or which have been misrepresented. The first impression given by the system to the users is also assessed at this stage.

The type of preliminary evaluation carried out is directly affected by the decision whether or not to use rapid prototyping for system development. Therefore, the case when rapid prototyping is used and, when it is not are treated separately in this section. However, the overall aims of the preliminary evaluation stage are the same, regardless of the approach used.

In both cases, whether or not rapid prototyping has been used, this stage of the evaluation is carried out by the experts who were used to develop the knowledge base. The introduction of new reviewers comes at a later stage.

If Rapid Prototyping has not Been Used

If rapid prototyping has not been used, preliminary evaluation is generally the stage of the evaluation which takes place immediately after the first prototype has been developed. The stage is equivalent to the final stage of the KE methodology: prototype analysis, which is described in Chapter 5.

If rapid prototyping is not used, the bulk of the knowledge elicitation is carried out before the initial prototype is built. Therefore, providing the Knowledge Elicitation has been done effectively, the first prototype will have been built from a substantial amount of the information required for a complete knowledge base and can be considered to be relatively accurate.

The preliminary evaluation involves sending the system to the experts who were originally used to develop it. Once they have seen the system and are satisfied with its content and the way it represents their reasoning processes, the system is allowed to progress to a more rigorous long-term evaluation.

When Rapid Prototyping has been Used

With this approach, criticism of the prototype is relied upon to elicit a large amount of the knowledge. Therefore, it is difficult to distinguish where KE ends and evaluation begins. It must therefore be assumed that the preliminary evaluation is occurring continuously throughout the KE, as the knowledge which is being put into the system is being evaluated by the expert(s) who are building it.

Long-Term Evaluation

This second stage of evaluation can begin once the preliminary evaluation is felt to have been completed.

When rapid prototyping has not been used for KE, this can be taken to be when the system satisfies the requirements of the experts who were originally used to develop it.

When rapid prototyping has been used, the changeover to long-term evaluation is less distinct and relies more heavily on the judgement of the developers. In this case, the developers have to decide when the system has reached a certain stage of development, and can be considered to be reasonably correct and complete.

This part of the evaluation aims to verify the accuracy of the knowledge base, assess the suitability of the developed system for the required industrial environment and ensure that the approach which the system adopts is actually the approach needed by the working environment in which the system is to be implemented.

The long-term evaluation also aims to investigate the directions in which the system could be developed in the future and establishes which other existing computer software the KBS could be sensibly linked to provide a more substantial overall system. Sustained evaluation such as this can help to identify the system's deficiencies, helping to distinguish how to improve its usefulness.

The long-term evaluation we have used and found to be effective involves implementing the system on a semi-practical basis, leaving it with a number of reviewers in their workplaces and asking them to give their impressions of it. This not only provides the opportunity for the reviewers to evaluate the information contained within the system in its normal

working environment, it also gives them time to consider when and how
the system would actually be used in practice. By leaving the system in
the workplace of the reviewers, the system is being evaluated in a familiar
and normal environment, as opposed to organised "trials", which would
seem unnatural to the user. This enables the applicability and deficiencies
of the system to be more naturally assessed, as the users can determine
when and how during their normal working day they would use the system. It also encourages more people to become involved in the system
evaluation: the benefits of which have already been discussed.

In both cases, the long-term evaluation should involve independent reviewers: that is, people who have not been involved in the system development before the evaluation stage. These can be new experts, users or
anyone else who would be interested in using the system. The benefits of
using reviewers such as these, who are new to the system, have already
been discussed.

The long-term evaluation can incorporate a number of additional
aspects, such as in-depth analysis of the user interface. In this book, the
user interface is not just taken to be what the user can see on the screen:
it also includes all the facilities which are available to the users via the
system (for instance, the glossary, explanation facility and so on). The
user interface is a very important part of a KBS as it is the part which
transmits and accepts information, and it is consequently vital for the
functionality of the system.

It has been our experience when developing KBS that the user interface
can greatly affect the user's opinion of a system, hence influencing their
confidence in it. In fact, when developing the bridge design system, the
user interface seemed to be so important that we began an associated
project which was specifically aimed at investigating the user interface
(see Chapter 9). This provided a number of interesting ideas and, overall it showed the great effect which a user interface can have on the users
of a KBS. An analysis of different interfaces and different knowledge
representation mechanisms can be very interesting and useful at this
stage, and helps to ensure that the developed system is well liked by its
users.

Aspects such as deskilling and interest level should also be taken into
consideration when developing the user interface. An interface which the
user feels is enhancing their abilities, as opposed to substituting for them,
is always preferable. Also, a system which maintains user interest helps
to ensure that the system is not rejected once it has been used a couple of
times. It was suggested by some of our reviewers that dividing the system
into sections would help to maintain user interest. However, this is a difficult thing to do with most KBS, as most of them do not deal with "sections" of the domain in any specific order: the order of the questioning
tends to be dictated by user responses and the way in which the knowledge base is searched. Therefore, even though we tried to section the
knowledge base during the evaluation, the same reviewers who suggested it decided that the sections were confusing. This indicates another
important part of evaluation: what users may think is going to be an im-
provement may not actually be so when it is included. Evaluation and
iterative development of the system enables them to see that not all their

criticisms are justified, hence helping to make them more receptive to the system as it stands.

User Interaction

The evaluation procedure should aim to assess the attitude of potential users to the system: a very important consideration in practical system development.

The accurate identification and involvement of potential users are very important in KBS development. In the past, user opinion has not been given sufficient emphasis, the focus of attention generally being given to the experts and their opinions (Kidd, 1985; Bramer, 1989). If the opinions and needs of the users are not taken into account, it is unlikely that the completed system will be accepted in its working environment. Similarly, as discussed in the previous section, if sufficient emphasis is not given to the creation of an effective user interface, it is unlikely that the system will be effective or fully utilised.

The only feasible way to develop such an interface is to collaborate in depth with actual users in order to discover their needs, opinions and requirements. The interviews which we have carried out with prospective users during the evaluation of the KBS that we have developed have proved to be particularly productive. During these meetings we discussed all of their views and suggestions for the system, which enabled us to orient the systems to their needs. Often it was found that the experts had a tainted view of what the users actually needed: either overestimating or underestimating their background knowledge and level of understanding. Also, the users are able to state areas of the system which they find confusing, and they can identify facilities which would help them use the system effectively.

One thing which should be taken into account when dealing with inexperienced users is that often this inexperience will be linked to lack of confidence. As the users tend to be interviewed in a group as opposed to individually, we have found that if they are not confident in their ability to contribute to the group discussion about the evaluation, they will often remain quiet as they do not feel qualified to comment. This can only be overcome by actively encouraging them to participate. Follow-up interviews, as described later, are a particularly good time for this.

Aims of the User Interaction

The aims of user interaction are fairly obvious: to ensure the content and aims of the system are acceptable to the users and that the chosen interface satisfies their requirements.

It also aims to elicit information concerning the future development of the system. This section of the evaluation can also help to highlight the differences in objectives and opinion between the users and the experts. This is important, as frequently experts do not fully appreciate user requirements.

Acquiring and Assessing the Evaluation Information

In any evaluation process, the way in which the information is acquired and recorded is of great importance. The following sections highlight approaches which we have used and have found to be effective.

The Diary

One approach is to provide all the reviewers with a blank book or "diary" in which they can record their comments each time they use the system. It should be emphasised when starting the evaluation that all comments are of importance whether they be on the content, appearance, interaction or future development of the system, and that no comment is too trivial.

We have found using a diary useful as it enables the reviewers to record their comments immediately, therefore placing no reliance on recall at follow-up interviews. This inevitably improves the reliability of the evaluation.

Recording comments in this way also helps the evaluators capture how a problem occurs as well as what it actually is. That is, the sequence of events prior to a bug, for instance, can be remembered and recorded, making that bug far easier to amend.

The diary also acts as a reference for both reviewers and system builders, helping the evaluators to recognise the comments that have been made by the users and to identify which have been included in the revised version of the system.

The diary system should be supported by a series of follow-up interviews, during which the diary and the comments contained within it can be discussed, and solutions to problems proposed.

Follow-Up Interviews

The follow-up interviews help the system builders to identify comments in the diary which are misleading or confusing. It also helps to familiarise the system builder with the opinions of the reviewers, helping to establish a rapport between the knowledge engineers and the evaluators, the importance of which has already been discussed in the knowledge elicitation sections. We have found that once this rapport has been established, the productivity of the interviews increases, and it is easier to understand the needs of the reviewers.

It is important to treat these interviews seriously as we have frequently found that they are useful for acquiring new information which has not been noted in the diary. Therefore, they are effectively acting not only as an evaluation process but also as extended knowledge elicitation. It is therefore important to ensure that they are successful.

We always tape record and transcribe the interviews in the same way as during the KE (see Chapter 5). This is important as many of the comments (particularly in group discussions) can be missed. Also, it enables the system developers to be free to take part in the discussions and not be distracted by having to take notes.

The interviews also help to verify that the criticisms are justified before they are carried out, and provide the opportunity to highlight conflicts between expert and user opinion.

Listing of Comments

The information contained within the diary and the additional information elicited during the interviews should be analysed and a list of criticisms and suggestions made. This numbered list of comments enables a clear assessment of the suggestions to be made and also helps compare different company's comments, enabling the effect of differing office environments to be assessed.

Once the required changes have been identified, they should be made and the new version of the system sent out to the reviewers so the whole process can begin again. Therefore, iteratively (hopefully!) improving versions of the system are produced. It is helpful to call the different versions of the system different names (e.g. A—, B—, etc.) as it helps to avoid confusion.

It is also helpful if successive versions of the system are delivered to the companies involved simultaneously, as this enables direct comparisons between their comments to be made and again helps to avoid confusion.

The Analysis of the Information Obtained

There are various methods available for analysing the information which is acquired during the evaluation process. In this section, we are going to describe an approach which we have used and have found to be effective.

Comments made about the system can be analysed and classified. Categories can include such things as:

1. Interface: These comments concern the user interface of the system: that is, comments on wording of questions, colour schemes and extra facilities such as glossaries and explanations.
2. Content: These comments are directed at the content of the system: that is, things which are felt to be missing from the knowledge base and mistakes which the system seems to make owing to lack of information.
3. Development: These are comments concerning the future development of the system: that is, suggestions of ways in which the system could be developed to be either more useful or acceptable.
4. Bugs: These comments are concerned with the programming of the system: that is, mistakes which the system makes which seem to be due to programming errors as opposed to gaps in the knowledge base.

There are obviously a number of other categories which can be used, and the types of category chosen will be influenced by the types of system being developed and also by the intended users.

Once the comments have been categorised, the number of comments in each category can be compared and the ratios of each set of comments obtained. This enables the comments to be compared in a number of ways. It also makes it easier to establish which comments have been made by more than one set of evaluators and which comments seem to be localised to one company.

The ratios of the comments can reveal a lot about the way the reviewers see the system. For example, when reviewing one system, we found that the large majority of the evaluators' comments were focused on the system's interface. Once the interface was improved their overall attitude to the system appeared to improve as well, indicating that the interface strongly effects the users' opinion of a system. This focus on the interface can cloud the reviewers' judgement, as they can only see faults or benefits of the interface as opposed to analysing the content or effectiveness of the system.

Finally, categorising the comments in this way makes a good record to refer to when developing the same or other systems, so the same mistakes will not be made again.

The Importance of Feedback in Evaluation

Feedback is an essential part of KBS development, both during the KE and the evaluation.

It is important that the reviewers feel that they are influencing the development of the system. They must feel that the comments they make are being taken into account and used, if they are to stay interested in the evaluation of the system. If changes which they suggest are not made for some reason, the reasons why these suggestions have been ignored must be carefully explained.

Feedback also helps to stimulate their interest and promote further co-operation. It is usually found that once the reviewers realise they are actually influencing the way the system is being developed, their enthusiasm increases and the quality of their comments improves, as they genuinely become interested in the way the system is improving.

Feedback can ensure that the reviewers' comments have been correctly interpreted. It is very easy to misinterpret a reviewers' comments and make the wrong adjustments to the system: by making sure that the reviewers see all of the changes which have been made to the system, this can be avoided.

We have often found, however, that even if the reviewers' comments have been correctly interpreted, once the requested changes have been made to a system, the reviewers no longer like the changes. This has been illustrated by the example given in the previous section, where, on the users' request, the system was divided into sections, but on subsequent review the users' decided that this sectioning was in fact detrimental. So, be warned! You may have to change things back again. This is particularly true when developing the interface. However, making the changes in the first instance is still worthwhile as it allows the users to see that the original approach is in fact preferable.

Summary and Conclusions

Basically, evaluation is a time-consuming and difficult process but, nonetheless, it is still a very important part of KBS development. It is important that the KBS assessment should incorporate all three of the aspects which have been identified, that is, evaluation, verification and validation, in order to ensure that the system is correct and useful.

Evaluation is particularly important in design domains because of the highly practical nature of the field, and it is vital to show that a KBS has been thoroughly tested if it is to be accepted by the average designer.

It is important to ensure that a broad range of evaluators are used, but it is vital to include potential users, as their opinions and those of the experts often differ greatly. Without their involvement it is highly unlikely that the system will ever be accepted and used in practice.

One very important aspect of successful evaluation is to be thorough and methodical. Keep track of the comments and versions of the system which have been made, and make sure you carry out the required changes within a reasonable amount of time. If this is not the case, you will soon lose the confidence and co-operation of your evaluators.

Overall, several approaches to KBS evaluation have been suggested in this chapter, but these are by no means rigid rules. In our experience, establishing a rapport with your reviewers is as important, if not more so, than in the KE process, so take the time and trouble to do this. Choose the approaches to suit your evaluators and, most of all, be flexible – you have to be able to fit in with the evaluators' requirements. You may have a vision of how you want your system to be, but it is the user and the evaluation process which should control the system development and not you! Only then can you be sure that the KBS will be effective.

Chapter 8

Examples of the Implementation of KBS for Conceptual Design

Objectives

This chapter aims to:

- Briefly outline the evolution of design KBS.
- Illustrate this evolution and the nature of design KBS using examples.
- Give three detailed case histories of design KBS which we have developed.
- Give readers some starting points and indicators on how to approach KBS development for design domains.

Introduction

So far we have looked at the background theory of both knowledge and design, and at the necessary techniques for applying these ideas in order to formulate a design KBS. However, a fundamental part of the human learning process is to study previous examples which use relevant techniques. Therefore, in this chapter we describe, in as much detail as space allows, the processes and thoughts which went into the creation of three KBS for conceptual design with which we have been involved. Two of the systems are for structural design and the third is in the domain of sewage sludge! As our background is Civil/Structural Engineering, the work is inevitably heavily biased towards this area. However, the descriptions are written in such a way that people interested in other areas of design will be able to see how a project develops, how the demands of the domain are coped with and, possibly most important of all, the way in which user needs are taken into account.

The description of each system covers the initial aims of the work, the knowledge acquisition, the structure of the knowledge base, the user interface, the chosen software and its features, and a description of a typical user session with the system. Obviously the latter is only a poor substitution for a real consultation, but nonetheless it helps to give an indication of the overall structure of the systems.

Before launching into a detailed description of these three systems, we will give a brief history of design KBS and an overview of some other design KBS which are described in the literature. This is done for three reasons: firstly, to illustrate the way in which design KBS have evolved, so helping to explain the changes which have occurred over recent years; secondly, because inevitably our own work does not cover the full extent of the available techniques for design KBS; and thirdly, to provide non-civil engineering readers with starting points from which they can obtain more information about their own area of specialisation. Obviously it is impossible to describe the work of others in the same detail as our own work.

Literature Review of Design KBS

Design is an activity which is undertaken by many classes of people ranging from graphic designers to software engineers. Although both ends of the spectrum involve creativity, in this book we are concentrating on the more technological aspects of design as opposed to the artistic (although we have included architecture which some consider to be artistic). This is reflected in our choice of examples of design KBS.

The Evolution of Design Knowledge-Based Systems

As has already been discussed, the roots of KBS lie in diagnostic systems, with such subjects as medical and chemical diagnosis being the domains which were originally tackled (Feigenbaum et al., 1972; Shortliffe, 1976).

Design oriented KBS were not seriously considered viable for a considerable length of time after diagnostic systems were instigated. For many years, design was felt to be an "inappropriate" domain for the development of such KBS: it was considered impossible to build an effective system in this area. There are a number of possible explanations for this: the established familiarity with diagnostic systems, fear of the unknown, people being discouraged by the "black art" reputation which design seems to hold and also the difficulty in obtaining the relevant design information to encapsulate in the system. It is undeniably true that design oriented systems are one of the most difficult types of system to develop. However, some of the anticipated problems were because people were comparing them to diagnostic systems and therefore tried to build design systems which operated in the same way. The two areas are obviously very different, but as the only technology which was available was developed for diagnostic KBS, there seemed to be no other route to follow.

Although it is difficult to say with certainty when the first design KBS was developed, an early example was the SACON system developed by Bennett and Englemore (1979). This illustrates the above discussion, showing that this system did indeed try to emulate a diagnostic system. SACON used the EMYCIN shell developed from the pioneering medical consultation system MYCIN (Shortliffe, 1976). The purpose of SACON was to provide advice on the choice of a suitable analysis strategy for a given structural analysis problem. The system was linked to a large finite element based structural analysis program. The complexity of the latter was such that to learn how to use the program in an appropriate manner required a lot of experience, and so SACON was developed to try and provide a source of expert advice.

Therefore, it can be seen that this system was essentially a diagnostic design system: that is, it chose an analysis strategy, so using the same techniques that had been developed for diagnostic systems. This example also illustrates another characteristic of early design KBS: they tended to concentrate on the more tangible areas of the design process, that is, design analysis. Analysis was an area of design in which effective computer systems had already been developed and applied, and where computer implementation was recognised as being beneficial. At this time, conceptual design was still a stage of the design process where no previous computer systems had been developed. Consequently, KBS for this area were not considered.

In the early 1980s there was then a gradual increase in the work on the development of design KBS, for example, the work of Brown and Chandrasekaran (1983), Brown (1984) and Maher (1984). About this time came the introduction of KBS which encoded such things as Codes of Practice, enabling them to be "intelligently" searched. These KBS were developed to replace the existing clumsy literature, therefore essentially acting as more effective text books (Kumar and Topping, 1989).

From here, work eventually progressed onto "real" design, with developers beginning to look at the preliminary stage of the design process (Maher et al., 1987; Gero and Maher, 1988). For a summary of much of the remainder of this early work the reader is referred to Pham (1988) which contains details of a large number of KBS. Also for Civil Engineers there are some descriptions of design KBS in Adeli (1988a).

From 1984 onwards there was a steady increase in the number of design KBS which were being developed. One of these, the work of Soh (1986), is reported below. However, despite the fact that these KBS were dealing with a completely new area, they still tended to treat the information in a similar way to diagnostic systems: using similar software to develop the systems and similar formats for both the user interfaces and the solutions which were presented. This severely restricted the effectiveness of the systems which were developed, and inevitably controlled the amount and type of information which could be successfully represented and manipulated.

From this, researchers and developers alike began to realise that possibly this was not the best way forward for design KBS: if they were to succeed, new approaches would be needed. A section of the research field moved over to investigate intelligent database design, as they recognised the space and storage difficulties which had been encountered with previous design systems. In addition, the importance of previous designs in the overall design process began to be realised. Indeed, one of the examples given in this chapter illustrates this emphasis on co-operating database technology (Soh, 1986).

Others started to investigate product models for the design process, with the view that once these models had been established, building successful KBS for design would become more feasible.

However, some researchers, such as ourselves, still feel that there is a place for KBS, but these KBS have to be built to complement human expertise as opposed to supplanting it. The future of KBS (as we see it!) is discussed more fully in the following chapter. It is sufficient to say here that in recent years the attitude of KBS developers to design KBS has changed enormously. They now realise that steps have to be taken to recognise design KBS as a subject in its own right, and one which needs to be dealt with differently from the existing diagnostic systems which essentially are selection systems.

The detailed examples of our work which are included in the chapter help to illustrate the later stages of this evolutionary process in terms of our own work: with the three systems developing to give an improved way of representing the design domain.

Some Recent Examples of Design KBS

As has already been mentioned, an important part of gaining experience in any area is to review previous examples. In the following sections we will look at a few examples from the literature so that a broader range of examples are covered. These examples concentrate on the later stages of the evolutionary process, as we feel this is where the design systems which are being developed are beginning to be of real use. Inevitably we have omitted much, including the work of some very eminent people. The amount of published work is so great that it is impossible to cover everything and so we apologise to those whose work is omitted, but our choices are based on a desire to show a range of design KBS.

For those who wish to know more about particular areas of KBS development, the references section includes work which is not covered in the text.

Example 1: A Progressive System

We will begin with an example from architecture. In Chapter 4 we covered the difficulties associated with cognitive overload and the problems faced by designers because of new materials and increasing complexity. Architects are by no means immune from these problems. The increasing pace of technological change and the ever expanding body of regulations and design codes mean that without computer-based assistance the architect's job would become increasingly difficult. In an attempt to provide a tool to help with these problems, a research team at Karlsruhe University has been working on the development of the Armilla system (Gauchel et al., 1990; Drach et al., 1992).

Armilla provides a toolbox which is implemented on workstations. The applications of the system which have been developed relate to the conceptual design of service-system layouts in pre-fabricated system buildings.

The software includes extensive graphics capabilities which give an on-screen view of the progress that the system has made. In the provision of services section, the system finds routes for a given service through the building and, as it makes decisions, these decisions appear on the screen. The routing process has to include a good understanding of the physical reality of the building such as where the columns and floor slabs are placed and what the intervening spaces represent (this is not as easy as it might sound). Allowances also have to be made for other services. For instance, if the air conditioning ducts are the first service that the system considers, then these must be installed in a way which allows the other services to be fitted in a sensible manner. This involves a lot of heuristic reasoning.

The work on this system started in 1985 and has suffered from a common problem in KBS development: that is, the rule bases becoming very complex and unwieldy. Problems have also been experienced with knowledge acquisition. In response to this, the system has been made more accessible to designers so that they can extend and change the system. This feature seems to be compatible with the way in which architects work, as the design knowledge which they use is not static, but instead it changes and evolves as the design progresses (as is true of design in any discipline). As with other design KBS, the work to date has concentrated on standard situations and solutions. Coping with anomalies and the unexpected is difficult with current technology.

The Armilla software uses the concept of dividing the design into stages in an attempt to reduce the degree of complexity. For example, for service systems five design stages are used. These are:

1. Spatial organisation of the building structure.
2. Requirement definition for service systems.
3. Strategic layout design of service systems.
4. Spatial co-ordination of duct/pipe/cable routes.
5. Component specification and allocation.

The knowledge structure of the system is too complex to describe here. A good description is given in Drach et al., 1992.

Interestingly, in Armilla, the more recent version of the system, the system is moving away from a pure expert system approach to a more interactive form of design system, where the user is allowed a far greater degree of input into the design process with the system providing warnings and checks only when required. We feel that this is a sensible route to follow. As stated by Drach et al., 1992:

"We do not think that tools [i.e. design KBS] should replace human control and communication but could be quite efficient in facilitating it."

Example 2: The Benefits of Using a Restricted Domain

McCarthy and Nouas (1991) have developed a system using the GOLD-WORKS toolkit to help designers with the detailing of structural steelwork connections. The domain is relatively small and well defined, which has enabled a substantially complete design KBS to be constructed.

The system has been developed using a combination of knowledge sources, which have included fabricators and in-house research experts. We have included this work here because it shows that if the domain is small and well defined, it is possible to create a very useful design aid using relatively simple and readily available software tools, albeit for a specialist area. Although many structural designers have an extensive knowledge of structural steelwork connections, there is always scope for providing the designer with a tool which will help with unusual or difficult situations. Such systems can be run as stand-alone applications and are of immediate use, whereas some of the more complex design KBS will only begin to be of real benefit when they can be linked to other design software. In addition, small scale systems such as this which are concentrated on specific areas are more likely to be accepted in practice than larger systems which claim to cover much wider domains. This is because people are more likely to believe that a system is capable of adequately covering a small domain.

Example 3: Systems Using Co-operating Agents

Morse and Hendrikson (1991) describe an object oriented blackboard system which uses the idea of co-operating agents. This means that the system consists of several knowledge bases, each of which covers a particular area of expertise, and which can interact with the other knowledge bases. With this type of system, obviously conflict resolution between the different agents plays a vital part in arriving at a design solution which is acceptable to all concerned. Conflict resolution is handled by a section of the software which together with the assistance from the agents formulates feasible alternatives.

The system has been implemented using the domain of the design of steel flooring framework systems to support manufacturing facilities. This is a fairly modest design task which is currently usually undertaken in the following sequence: placement of mechanical equipment for manufacturing efficiency, configuration of steel floor framing system, structural analysis and finally constructability assessment. Morse and Hendrikson (1991) argue that if the design stages are undertaken primarily in parallel as opposed to in the above sequence then the final design will be improved. In their system, each of the co-operating agents represents one of the above processes and this approach allows this parallel design to take place. As yet there is no proof to substantiate their arguments but the work does show that the use of co-operating KBS does allow one to restructure the design process, if this is advantageous.

The use of co-operating agents for mechanical engineering design has been investigated by Huang (1990) who based the architecture of his system on the work of Bond (1988). The domain of application is the design of machine tools.

A similar system to the previous two examples is described by Phillips et al. (1989). This time the domain of application is aircraft engines. When undertaken manually, aircraft engine design is a lengthy iterative process involving a large team of people who possess various specialities. A KBS was developed to undertake the design process automatically. Large time savings are reported although the report only covers a prototype system specifically used to establish the principle.

As yet, systems employing co-operating design agents have not reached the state of maturity where one can categorically state that for certain types of design problem they offer great advantages. However, potentially the technique is very useful. The development of such systems is more difficult than conventional KBS because of the additional problem of conflicts between the design agents.

Example 4: Case-Based Reasoning

Much of the design carried out in practice relies on the use of previous designs. Often, when a new design task is started the first reaction is to look at past designs of similar products. These designs contain a wealth of information which designers can use as a good starting point for their new project. This inevitably saves a great deal of time and is therefore an economic approach. Therefore, although design is a complex task, much of the design carried out in practice is not of the "open-ended" creative type but of the relatively routine redesign type. From the point of view of automating design, routine redesign is a significantly less complex task than creative design. However, it is not a trivial task – the designer must still evaluate the previous design and determine how to modify it in order to match the requirements.

For human designers, the major drawback of using past designs is that they find it difficult to keep track of previous work, owing to such things as forgetfulness and poor filing systems. Often there are no official records of where specific designs are kept, and consequently locating and using them relies predominantly on human recall.

Commercial Computer Aided Design (CAD) systems are widely used in industry. However, in general these can only help with the drafting stage of the design process. That is, they provide powerful tools which aid the creation and modification of the graphical representation of a design, and these systems are primarily concerned with the representation of points, lines and areas of space. Although they are becoming more flexible (for instance, many systems now allow the annotation of design), their facilities for processing information are limited. These systems are therefore incapable of supporting many aspects of the design process including verification, design critiquing and analysis. Much of the responsibility for selection of designs and the necessary interpretation is taken by the human designer.

Routine redesign has attracted the greatest amount of interest among AI researchers and this interest has motivated the development of such methods as constraint propagation and plan-based techniques. In addition, people have been investigating the capabilities of so-called "Case-Based Reasoning" (Dyer et al., 1986; Ashley and Rissland, 1988; Goel and Chandrasekaran, 1989; Maher and Zhang, 1991). "Case-Based Reasoning" is the name given to using AI techniques to keep previous design information as "cases" (i.e. in a standard format) which can be easily stored and manipulated. More importantly, storing previous designs in this way enables new and amended cases to be easily added to the existing database, hence over time creating a substantial and useful cache of design information. The main benefit of this approach is that a computer system such as this will never forget a case which is saved in its database, consequently overcoming the problem of relying on human recall. Similarly, if these cases are appropriately indexed then the computer system will not have any problems in retrieving cases, which is in direct contrast with the filing systems which are currently in operation in some design offices. Also, by storing designs which are failures as well as those which are successes, the designer can be made more aware of certain design deficiencies and is therefore more able to anticipate and avoid future design failures. Thus the computer system can be used to augment the human expert's memory by providing appropriate past designs, while still allowing the human to use his/her superior skills of recognition, invention and creativity to adapt the retrieved cases. Therefore the computer is used for what it does best: effectively storing and remembering large amounts of information while maintaining the important aspects of human activity and creativity. This counteracts the deskilling argument put forward by many (see Chapter 1) and also contests the theory that automating the design process would lead to the degradation of the design process as a whole (Colgan and Spence, 1991; Ishii and Hornberger, 1991). For a more detailed discussion on the subject of deskilling, see Chapter 1.

The representation of designs in a case database raises issues associated with how to represent cases, how to index them and how to select suitable cases from the stored database. These are considered briefly below, but for a more detailed discussion see Kolodner (1991).

1. Representing Cases: There are a number of ways in which a design case can be represented, including as a whole design, as systems, as systems referencing subassemblies and as design plans.
2. Indexing Cases: This is probably the most important issue in CBR, as it is through the chosen method of indexing that suitable cases are selected. That is, by assigning a label to the design when it is stored, it can be retrieved effectively when needed.
3. Case Selection: Even when appropriate indices have been selected, the process of identifying suitable cases still exists. This is complicated by the fact that generally a partial match will be searched for as opposed to a total one. A number of approaches have been developed to cope with this issue, including the use of heuristic metrics, algorithms for determining each case numerically (Pu and Reshberger, 1991) and methods which take past experience into account (Ashley and Rissland, 1988). Another method which has been developed is concept refinement which relies on the case database being organised hierarchically, in which the top hierarchy is more general and the lower levels of the hierarchy are more specific. Therefore, the search travels down the hierarchy "matching" nodes to the search criteria. Thus cases can only be retrieved when their abstractions match the necessary requirements.

The viability of Case-Based Reasoning as a practical approach has been shown in a number of research projects (Dyer et al., 1986; Hammond, 1986; Maher and Zhang, 1991). However, as yet the ideas have not been commercially exploited. With continuing development, it can be predicted that this is a technique which is ripe for deployment in commercial applications and which will prove to be extremely useful for the capturing and manipulation of past design experience.

Detailed Applications

The previous section contained recent examples of different design systems which have been developed by other researchers. This section contains detailed case studies of three design KBS with which we have been involved. These examples are presented in the order in which they were developed and it is hoped that the three examples will exhibit the same evolutionary qualities that were discussed in the previous section; that is, the principles behind the system development changing as more experience is gained by the developers and as the three systems develop.

Case Study 1: A KBS for the Preliminary Design of Offshore Structures

In this section we will cover the development of a KBS for the design of fixed steel offshore structures, which are typically known to the layman as oil rigs.

Introduction

The oil industry has for several decades been extracting both oil and gas from beneath the ocean bed. During this time a mature field of design knowledge about offshore structures has been developed, particularly for "shallow" water structures (shallow here being defined as not more than 100 metres). Deeper water environments such as the North Sea between Britain and Norway are also now being exploited and both concrete and steel structures have been built, but typically for shallow water environments steel structures predominate.

These structures are said to be "fixed", because although they are built away from their intended place of use, they are then towed out and fixed into position. The fixing is achieved using steel piles inserted through the legs of the lower part of the structure. A schematic layout of a typical structure is given in Fig. 8.1 which shows that the lower part of the structure, which is largely submerged, is called a jacket and the upper part is called a topside.

The domain of the system which forms the subject of this section is shallow water fixed steel structures. The work commenced in 1985 at MIT in the USA. The developer was Soh (1986) who worked under the supervision of Professor J. Connor and their collaboration produced a prototype system for the design of the jacket structure. Soh (1990) subsequently undertook further research at Cardiff University and extended the system to include the design of both the topside and the jacket, as well as coupling the KBS to a graphical editor and a finite element analysis package. Soh had previously worked on the design of offshore struc-

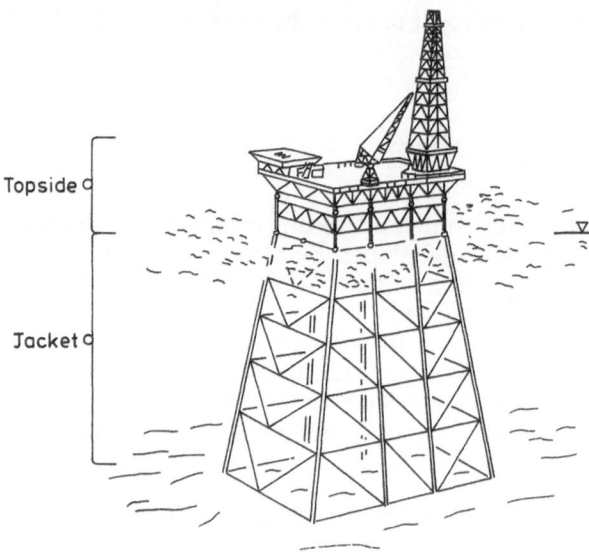

Fig. 8.1   A fixed steel offshore structure

tures for the South East Asia region and the system is based on his knowledge of the design of such structures.

The system is PC-based running under MS-DOS and although it will operate on a 286 machine, more acceptable response times are obtained from the faster PCs. The system is written in a mixture of dBaseIII, Clipper and C, with the graphical editor being in GKS. The original system (Soh, 1986) was written using GEPSE which was a shell developed at MIT (Chehayeb and Connor, 1985).

The initial project aim was to develop a prototype system to test the applicability of KBS techniques to offshore design. The later work extended this concept to look at the utility of coupling KBS to other types of software. To date the system has not been implemented in a practical environment.

During the development of any system, even if it is never to go past the prototype stage, it is necessary to have in mind potential users. This system was designed to be of help to relatively inexperienced design engineers (say with less than 5 years' experience) who traditionally have at various stages of the design process to call on more senior engineers for help with such things as the topology of the structure and preliminary sizing of members.

Finally, before moving on to look at things in more detail, no computer system is truly complete without an acronym – in this case it is IPDOS (Interactive Preliminary Design of Offshore Structures).

Choice of Domain and Aims of the System

For IPDOS the choice of the domain was dictated by the main developer's previous experience in the Offshore Industry of South East Asia. The system developer also acted as the domain expert. Given the problems discussed in Chapter 2 about procedural knowledge and the difficulties of converting such knowledge into a form where it can be used in a computer system and explained to others, C. K. Soh's achievement in building this system is very impressive. Some help was obtained from relatives and friends but the bulk of the knowledge base has been developed by Soh himself.

The aim of the original version of the system (Soh, 1986) was " . . . to explore the possibility of applying [Knowledge Based] systems to offshore engineering" with the more specific aim of designing and implementing a prototype KBS to assist engineers with the selection of the basic structural configuration during the conceptual design of fixed steel offshore structures. So the system would suggest a suitable form for the arrangement and spacing of the members of a jacket structure but would not pick any section sizes or design the topside. Following the pioneering work of Maher (1984) the original system represents an early application of KBS technology to design in structural engineering.

The further work on the system was undertaken first to extend the domain of application to include the topside and also to determine preliminary member sizes and undertake code checks. A further component was added to the system at this stage, this being a graphic editor which enabled the structural models generated by the KBS to be examined visually and, if necessary, members inserted and removed. The coupled KBS/graphic editor can also be used to produce a file which can act as input data to a suitable analysis package. For IPDOS the system was coupled to the STRUDL (Soh, 1985) package but any other suitably powerful analysis package could be used.

The coupling of KBS/graphic editor/analysis package was introduced because it was recognised that as a stand-alone tool a KBS is significantly less useful than when it is coupled to other software with complementary abilities, so that in this case almost the entire conceptual and embodiment design process and the initial part of the analysis could be achieved using the coupled system.

Software and Hardware

The development of IPDOS began using the GEPSE software tool (Chehayeb and Connor, 1985), but subsequently the system was rewritten using dBaseIII as the main language with knowledge-based additions written in Clipper (a compilable language which is compatible with dBaseIII) and C. The resulting suite of software tools has the advantage of combining the power of a relational database with knowledge-based programming techniques. The whole toolkit is called KBASE (Soh, 1990) and although the software was specifically developed for this project, it potentially could be applied to a wide range of problems/domains. To date the software has only been implemented on PCs running under MS-DOS. Many problems have been experienced because of the 640K limit on system size which is imposed by the DOS operating system. If the system is to be enhanced then either DOS extender technology will have to be used or a more powerful workstation version will need to be developed.

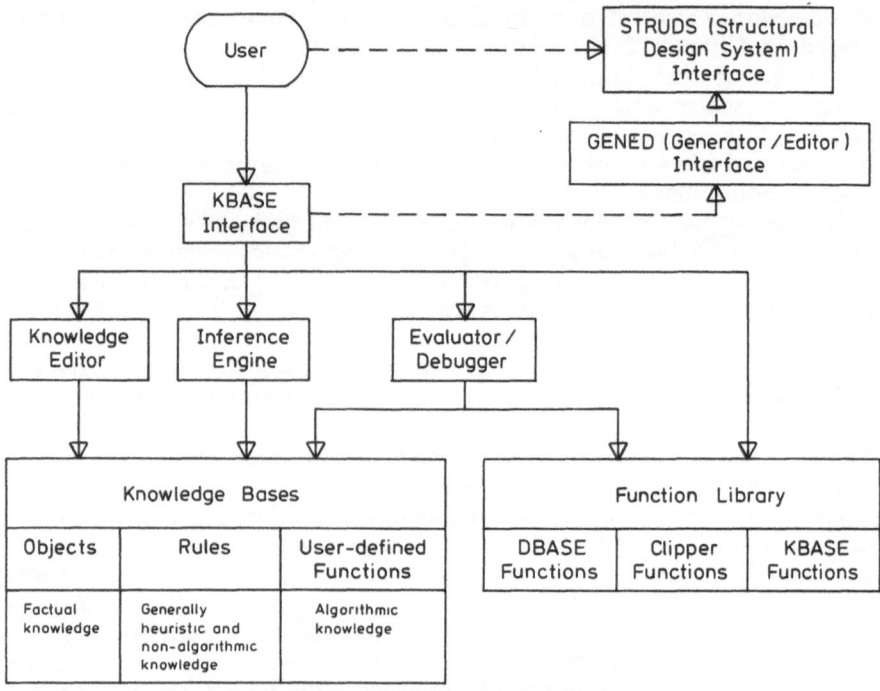

Fig. 8.2   KBASE: system architecture

The architecture of the system is given in Fig. 8.2. It can be seen that KBASE uses a hybrid form of knowledge representation which consists of an object-oriented knowledge base with rules acting on the objects. Also the knowledge contains user-defined functions which are used to store algorithmic knowledge.

Each object within the knowledge base is represented as a frame with an unordered list of attributes which then have associated values. An example of such a frame is:

"PLATFORM" object
Jacket: jacket
Function: wellhead
Water—depth: 300
Topside: topside

The water depth is given in feet because the offshore industry tends to dominated by the USA (1 foot = 0.3048 metres).

The words before the colons are field names and the words after are the values. The "JACKET" field takes the value jacket which happens to be an object with the name jacket. At the moment there is no automatic inheritance between frames within KBASE, although the user can define an inheritance structure if required. Effectively the frames within IPDOS mostly contain declarative knowledge.

The heuristics are largely incorporated into the rule base of the system. The rules are expressed in the classical production rule form with some additions to cope with the requirements of dBaseIII, for example (Soh, 1990):

RULE SET : global
ASSIGNMENT : GET("Function","Platform") = "Wellhead"
IF : GETN("Water−depth","Platform")200.AND.assignment
THEN : PUT("Recom−inst−method", "jacket","launch−off").AND >
PROMPT("Crane may not have adequate lifting capacity.")

The GETN command is needed because dBaseIII stores all values as strings and so, where it is necessary to convert to a number, GETN (or PUTN or STR2NUM) has to be used.

The first line in the rule states which rule set the rule belongs to. KBASE allows rules to be arranged in sets (see Fig. 8.3 for the implementation of this in IPDOS). The upper set is called the global set and initially inferencing starts at this level and is then transferred as appropriate to one of the so-called sub-rule sets at the next level down (the next level down is then called a sub-sub-rule set etc.). The transfer procedure is controlled by meta-rules which are embedded within the rule sets. The reason for dividing the rules into rule sets is simply to increase the efficiency of the software by reducing the number of rules that have to be examined to solve a given problem.

The GETN command returns the value of the water depth, field of the object and platform, and then compares this with the value of 200 (feet). The ASSIGNMENT statement serves as a local variable and in this case appears in the predicate field (i.e. within the IF part of the rule). Thus if the rule succeeds and the water depth is more than 200 (feet) and the platform is to be a wellhead, then the recommended installation method for the jacket is to launch it from a barge because the size of such a structure will be too large for lift off using a crane. The latter is the preferred method where possible.

The rule is a good example of heuristic knowledge regarding what is best in terms of the construction process, the likely availability and cost of a suitable crane and the effect of the installation procedure on the form of the proposed structure. If the jacket is to be launched, then one side of the jacket will have to be of a suitable form to allow it to slide from a barge.

The last component of the knowledge base is the functions. A function is similar to an object but with the following fixed set of fields: "PARAMETERS", "PRIVATE", "BODY", "RETURN". The "PARAMETERS" field contains the values which are passed to the called function and "PRIVATE" is the declaration of local variables used within the body of the function. "BODY" contains a series of dBase commands and other functions to be executed, and "RETURN" contains the values of the variable or expression which the function returns. For example:

"DXN1" function
PARAMETERS: jacket, leg—batter
PRIVATE: bax—c
BODY: bax—c = GETN("BAX—C",leg—batter)
RETURN:
IIF(bax—c0,GETN("Height",jacket)/bax—c,0)

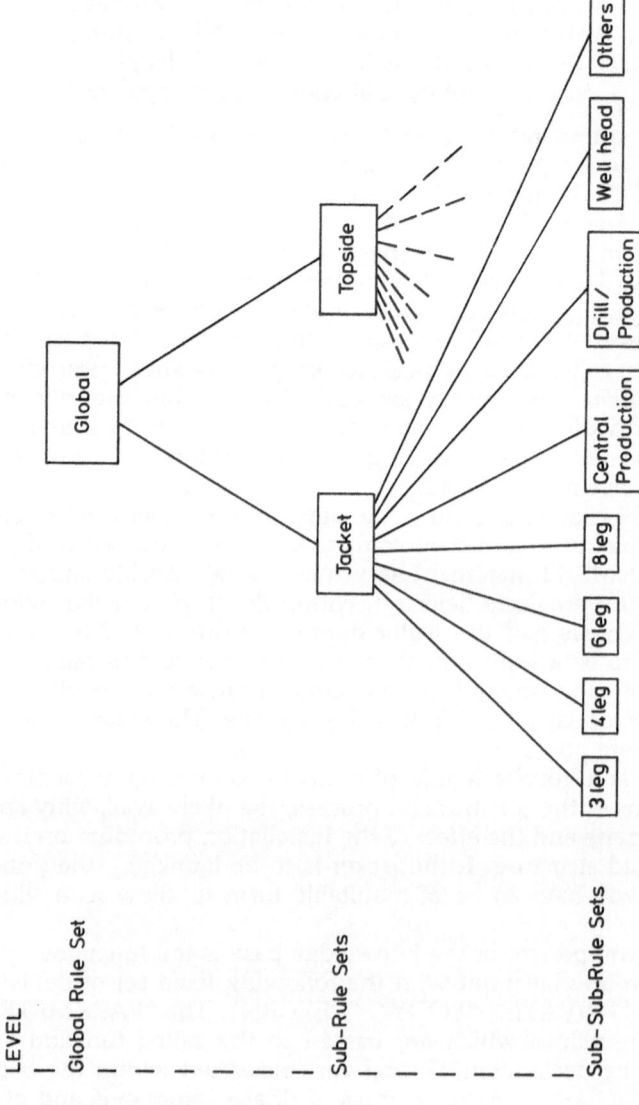

Fig. 8.3 IPDOS: rule set organisation

LEVEL

Global Rule Set

Sub-Rule Sets

Sub-Sub-Rule Sets

In this example the function DXN1 is passed jacket and leg—batter as arguments. Its body then retrieves from the IPDOS object base the inferred x-direction batter of the jacket leg C and assigns the retrieved value to "bax—c", a local variable of the function DXN1. If "bax—c" is positive, DXN1 returns the height of the jacket divided by "bax—c", otherwise it returns a value of zero. KBASE uses the IIF statement in exactly the same way as IF. The IIF statement is applied within user-defined functions to avoid confusion with the production rules.

KBASE uses a forward chaining inferencing procedure, the process being controlled by a function which is appropriately called INFER.

System Features

The features that occur in the IPDOS system do so because of the capabilities of the KBASE software. Thus in this section we will use examples from IPDOS to show the capabilities of KBASE.

The system works using the questioning procedure which is common to almost all those KBS which can also be classed as expert systems. The questions appear at the top of the screen (see Fig. 8.4) and where ever possible the user reply is restricted to pre-determined answers which are chosen from a menu. The choice is made by using the up and down arrow keys to move the highlight bar to the preferred answer and then by hitting the Return (Enter) key. Where free form input is allowed, such as when the user is asked the depth of the water at the installation site of

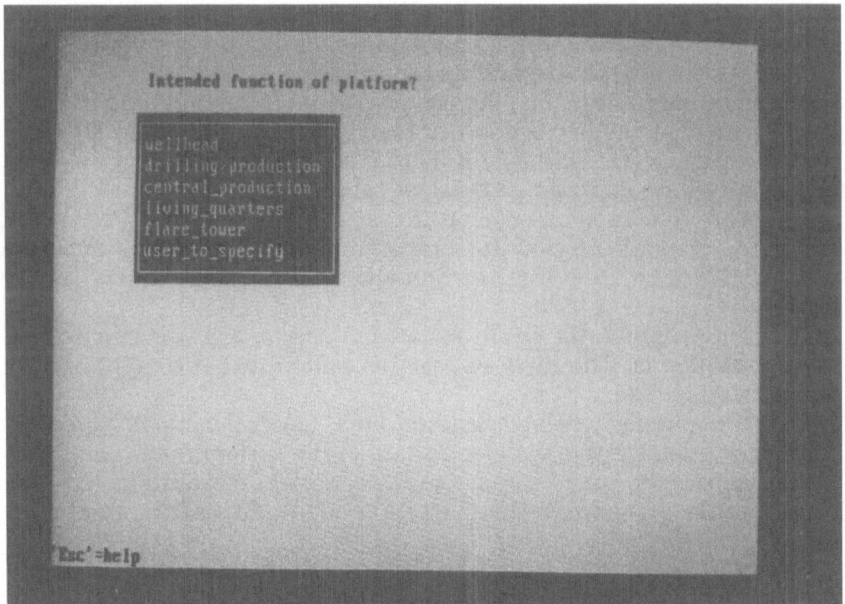

Fig. 8.4 Question format for IPDOS: a typical screen

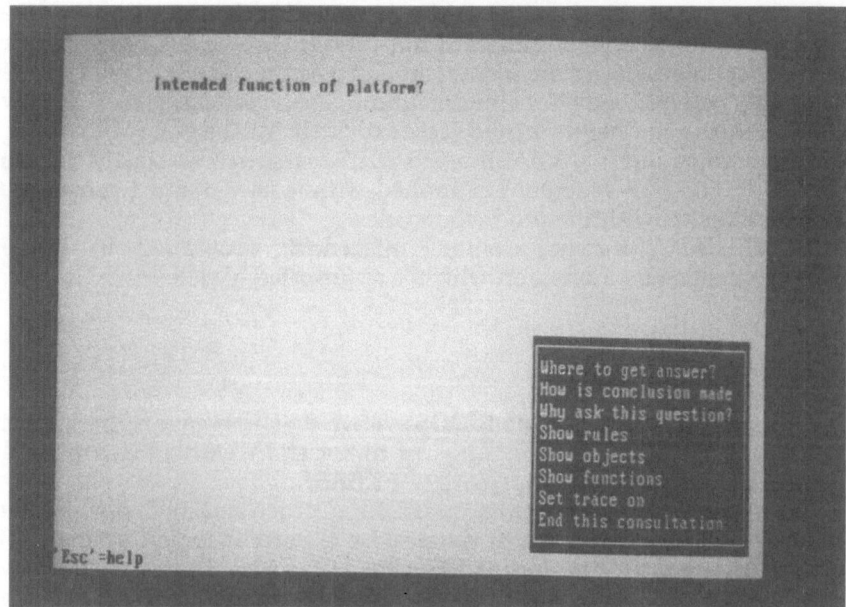

Fig. 8.5   Help system of IPDOS: a typical screen

the proposed platform, then the system checks the answer to ensure that it is not outside the values for which the knowledge base is valid and that it is sensible (e.g. a water depth of 4 feet would not require a normal offshore structure).

The restriction of input in this way ensures that the number of possible errors in input data is kept to a minimum and that the system is not used for problems which are beyond its capabilities.

As can be seen in Fig. 8.4, the user can at any time press the ESC key to obtain help. The help screen which then appears is shown in Fig. 8.5. So the user can ask for assistance if there is any uncertainty as to where the requested information can be obtained, there is a facility called "Where to get answer". The reply appears as text at the bottom of the screen. Below this in the menu is a facility for explaining how the present conclusion is reached. Within KBASE this function gives a list of rules which have been fired.

The next facility in the menu allows the user to ask what the relevance of the question is. The reply appears as potted text at the bottom of the screen (see Fig. 8.6).

The "Show Rules", "Show Objects" and "Show Functions" items allow the user to examine the knowledge base of the system. The "Set trace on" feature allows the user to trace the inferencing procedure as the system runs, and the last item in the menu is to allow the user to abort a given run.

Once the initial questioning procedure is over, the inferencing proce-dure begins. During the inferencing procedure IPDOS provides extra in-formation to the user to prompt thoughts about some features of the sys-

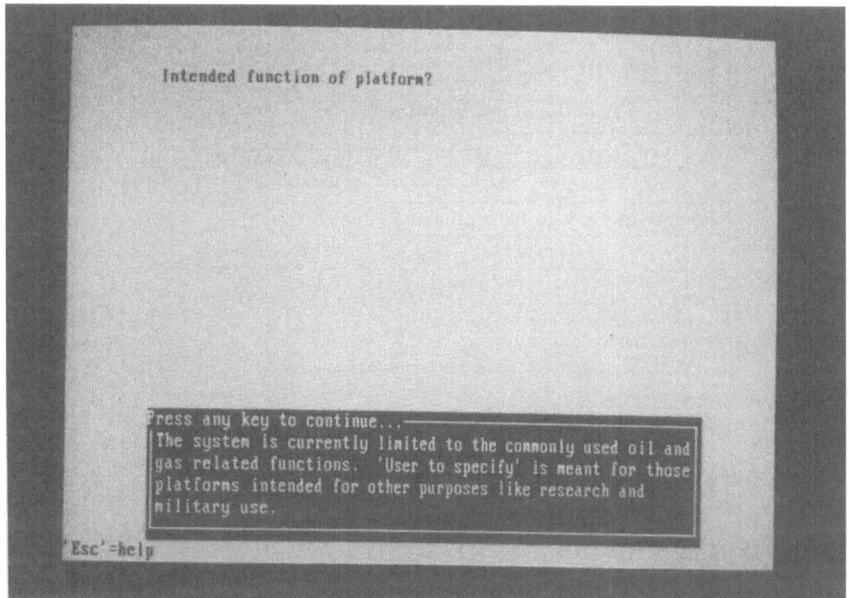

Fig. 8.6   Output text from IPDOS help system: a typical screen

tem which might not have been considered and also to provide information regarding the progress of the inferencing procedure. This is part of the teaching facility of the program, passing on knowledge to inexperienced users, and also during the inferencing which takes up to a minute, the displaying of messages helps to inform the user that the system is still active. At the end of the inferencing procedure, IPDOS produces a structural model of the proposed structure. This is then passed to the graphical editor which is used to display the structure, and the user is then offered the facility to change any of the recommendations of the system.

An Example of a System Consultation

To further help with the explanation of the features of IPDOS, an example of the use of the system is given below. The description will focus on the KBS part of the system with only a brief description being given of the graphical editor and the coupling to the analysis software.

Upon starting the system, a function which is appropriately named "First—Function" is called. This function first displays a graphical introduction screen and then subsequently handles the asking of the questions that the system uses to gain sufficient information to solve the particular design problem. Examples of the question format have already been given in Figs. 8.4 to 8.6.

The logic of the question procedure for the Jacket part of the structure is given in Fig. 8.7 (Soh, 1990). A similar arrangement is used for the top-

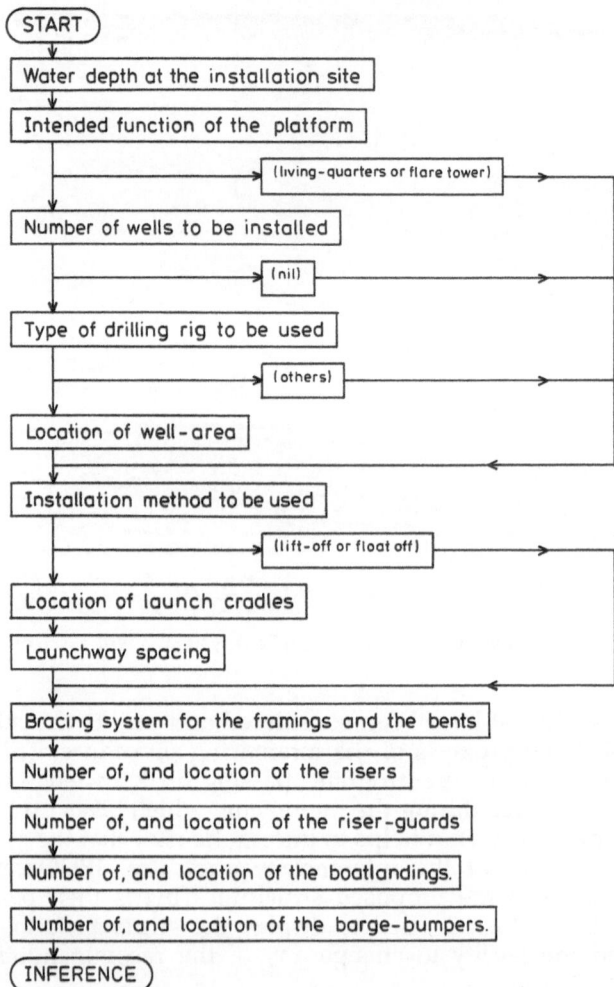

Fig. 8.7 Question procedure logic for jacket design: IPDOS

side design. As explained above, the user is able to access the help facility if assistance is required.

When the question procedure is completed, the inference process is activated using a function called INFER. If required, the user can access the help facility to activate the trace facility. This then enables the rules fired during inferencing to be identified. For a jacket design, examples of some of the fired rules/functions are given in Fig. 8.8. The notation is given in Fig. 8.9. Note that rules such as 4—4, which need to use the re-commendations inferred by other rules, typically start off by checking if the user has altered the inferred values. In rule 4—4 this is achieved by using the statement !KNOWN("Change","leg—batter") to check for changes in the inferred batter of the jacket legs.

As KBASE uses forward chaining, new values are inferred as the inference progresses. This makes it possible for rules which were previously triggered (i.e. the IF part partially satisfied) to be fired in subsequent passes over the rule base. For example, rule META—4LEG is only triggered in the round before the rule WELL—8 is fired. After WELL—8 is fired, it becomes known that the jacket is to have 4 legs, then all the IF part of the rule META—4LEG is satisfied. Hence this rule is fired in the next round

RULE	:	4_4
Rule set	:	*4_LEGGED*
If	:	*!KNOWN("Change","leg_batter").AND.*
		GET("Location","launch_cradle")="jacket_leg".AND.
		GETN("Spacing","launch_cradle")=GETN("TY","jacket_top")
Then	:	*THEN4_4()*

FUNCTION	:	*THEN4_4*
Body	:	*PUT("BAY_A1","leg_batter","100000")*
		PUT("BAX_A1","leg_batter","8")
	:	*PUT("BAY_A2","leg_batter","100000")*
		PUT("BAX_A2","leg_batter","8")
	:	*PUT("BAY_B1","leg_batter","100000")*
		PUT("BAX_B1","leg_batter","8")
	:	*PUT("BAY_B2","leg_batter","100000")*
		PUT("BAX_B2","leg_batter","8")
		PROMPT("Need to check the vertical bent bracing for dynamically"+;
		"induced forces due to the launching operation.")

RULE	:	4_7
Rule set	:	*4_LEGGED*
If	:	*GETN("Water_depth","platform")1>=110.AND*
		GETN("Water_depth","platform")<=240
Then	:	*PUTN("Number","horizontal",INT((GETN("Height","jacket")-+;*
		110)/40+4)).AND.;
		PUT("Recommended","bent_brace","combine_K_diagonal")..AND.;
		PROMPT("Use diagonal-bracing for top 100ft & K-bracing+;
		for the rest. Same for both bents and panels.")

RULE	:	4_9
Rule set	:	*4_LEGGED*
If	:	*KNOWN("BAX_A2","leg_batter")*
Then	:	*THEN4_9()*

FUNCTION	:	*THEN4_9*
Body	:	*PUTN("BY","jacket_bottom",;*
		DYE3("jacket","leg_batter")+DYW3("jacket","leg_batter")+;
		GETN("TY","jacket_top"))
		PUTN("BX","jacket bottom",;
		DXN3("jacket","leg_batter")+DXS3("jacket","leg_batter")+;
		GETN("TX","jacket_top")
		PUT("Number","vertical","2")

Fig. 8.8 Example of some of the fired rules and processed functions for a typical design: IPDOS (*continued* on next page)

```
RULE         :    J22
Rule set     :    JACKET
If           :    !KNOWN("Modify_jacket","design").AND.;
                  KNOWN("Wave_height","platform").AND.;
                  KNOWN("Water_depth","platform")
Then         :    PUTN("Height","jacket",INT(GETN("Water_depth",+;
                  "platform")+GETN("Wave_height","platform")/2))


RULE         :    META_4LEG
Rule set     :    JACKET
If           :    GETN("Number","jacket_leg")=4.AND..NOT.GOAL_4LEG()
Then         :    GRAPHICS ("4legs.pcx").AND
                  INFER("4_LEGGED","GOAL_4LEGGED")


FUNCTION     :    GOAL_4LEG
Private      :    Mreturn
Body         :    mreturn = KNOWN("BAX_A1","leg_batter").AND.;
                  KNOWN("BAX_B1","leg_batter").AND.;
                  KNOWN("BAX_A2","leg_batter").AND.;
                  KNOWN("BAX_B2","leg_batter").AND.;
                  KNOWN("BX","jacket_bottom")
Return       :    mreturn


RULE         :    META_WELLHEAD
Rule set     :    JACKET
If           :    GET("Function","platform")="wellhead"
Then         :    INFER("WELLHEAD","KNOWN('TX','jacket_top')")


RULE         :    WELL_1
Rule set     :    WELLHEAD
Assignment   :    GET("Function","platform")="wellhead"
If           :    GETN("Spacing","launch_cradle")>=40.AND.
             :    GETN("Spacing","launch_cradle")<=50.AND.
                  GETN("Number","well")>=3.AND.GETN("Number","well")<=15.
                  AND.m-assignment
Then         :    PUT("TX","jacket_top",GET("Spacing","launch_cradle"));
                  AND.PUT("TX","jacket_top","45")


RULE         :    WELL_8
Rule set     :    WELLHEAD
If           :    GETN("Spacing","launch_cradle")=45.AND.;
                  GET("Function","platform")="wellhead".AND.;
                  GETN("Number","well")>=3.AND.GETN("Number","well")<=15
Then:        :    PUT("TX","jacket_top","40").AND.PUT("TY","jacket_top","45")
                  .AND.PUT("Number","jacket_leg","4")
```

Fig. 8.8   (*Continued*)

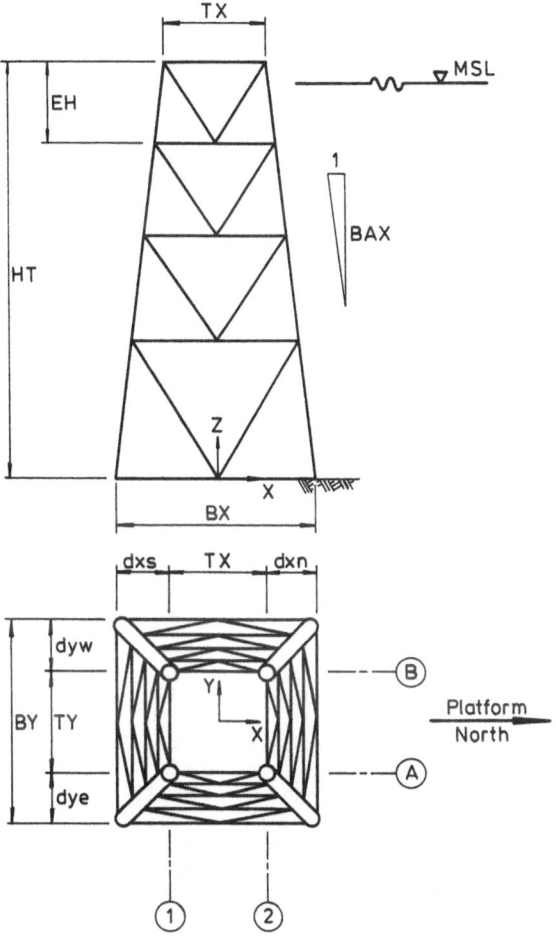

Fig. 8.9   Notation used for jacket structures

of inferencing. This results in the inference procedure transferring from the global rule set to the 4—leg rule set.

Once a rule has been fired and the functions which the rule accesses have been processed, the goals are tested to see if they have been satisfied. For example, in the case under consideration the inference engine looked at the rules in the 4—leg sub—rule set and the sub—sub—rule set wellhead before all the goals in the "Goal—Jacket" (not shown) and the "Goal—4leg" bases were satisfied.

When the inference procedure is finished the design information which has been input by the user is displayed so that it can be checked (Fig. 8.10) and the system then gives the proposed basic structural configuration (Fig. 8.11).

After the basic structural form has been inferred, the user is allowed to alter the system's recommendations. Three levels of alteration are permitted by the system:

DESIGN PARAMETERS AS INPUT:

Water depth at site (in feet)	>200
Function of the platform	>wellhead
Number of wells	>6
Type of drilling rig	>mobile_jack_up
Location of the well area	>next_to_bent_1
Installation method	>launch_off
Launch cradles are on	>jacket_leg
Launchway spacing (in feet)	>45
Horizontal frame bracing	>K_brace
Vertical bent bracing	>K_brace
Number of risers	>3
Location of risers	>A1/A2
Number of riser guards	>1
Location of riser guards	>A1/A2
Number of boatlandings	>1
Location of boatlandings	>B1/B2
Number of barge bumpers	>2
Location of barge bumpers	>B1_&_B2

Fig. 8.10   Input design parameters for typical design: IPDOS

PROPOSED BASIC STRUCTURAL CONFIGURATION:

Height of jacket (in feet)	>212
Length of jacket top (in feet)	>40
Width of jacket top (in feet)	>45
Number of Jacket legs	>4
Number of horizontal frames	>6
Number of vertical bents	>2
Length of jacket bottom (in feet)	>93.00
Width of jacket bottom (in feet)	>45.00
Recommended bent bracing	>combine_K_diagonal
Leg A1 batter in X direction	>8
Leg A1 batter in Y direction	>100000
Leg B1 batter in X direction	>8
Leg B1 batter in Y direction	>100000
Leg A2 batter in X direction	>8
Leg A2 batter in Y direction	>100000
Leg B2 batter in X direction	>8
Leg B2 batter in Y direction	>100000

Fig. 8.11   Basic structural configuration for typical design: IPDOS

1. Alteration with re-inference. This occurs when any alteration is of a magnitude such that it will affect other components. For example, when designing a jacket, any alteration in the height of the jacket affects the number of horizontal framings, the type of bracing for the vertical bents and the dimensions of the base of the jacket. Where such substantial alterations result from the alterations suggested by the user, the system advises that the best course of action is to re-run the consultation.

2. Alteration without reinference. This occurs when the alterations do not affect other components.

3. Alteration not permitted. Dimensions which are the end result of

component configurations, such as the dimensions of the jacket bottom and the spans of the topside deck legs, are not allowed to change.

When the user has completed the KBS part of the system the option of converting the KBS recommendations into a structural model using a function called GENERATOR is offered. The function is effectively part of the KBS software although not part of the main KBS. GENERATOR determines the topology and member incidences, and then undertakes the preliminary sizing of members.

Once the model has been generated it can be displayed using the graphic editor GENED. This allows the user to view the structure from any angle and members can be inserted or removed. Another feature allows parts of the structure to be viewed in detail.

Following on from the above, the system can then be coupled using a data file to any suitable numerical package so that the structural model can be analysed in detail.

Future Developments

At present IPDOS has usefully established the benefits of coupling KBS with other types of software but the coupling is unidirectional (i.e. from the KBS to the graphical editor and then to an analysis package). It would be useful if the knowledge base could interact with the other components of the system so that, for instance, when the first run with the analysis package is complete the knowledge base could form a useful aid with the interpretation of the results. At present, as with so much KBS research, this work is awaiting resources.

It would seem that the long-term prospects for linking software systems lie in the approach offered by product models in which either the internal representation of items is compatible between software systems or each system has the ability to interpret incoming information in a given format and to create output in the same format. If such flexibility becomes possible (work is underway via such initiatives as ISO/STEP and the European CIMSTEEL projects) then coupling of systems should become relatively easy.

Case Study 2: A Knowledge-Based System for the Conceptual Design of Road Bridges

In this section we will cover the development of a KBS for the conceptual design of road bridges.

Introduction

This KBS was always intended to be practically oriented and consequently it has been developed entirely in collaboration with industry. The domain covers small to medium span road bridges which cross another road. The system has been built using LPA PROLOG and is implemented on a PC, as this is the hardware which is most commonly available within engineering design offices.

The system is primarily intended for young civil engineers with no previous experience of conceptual bridge design. The system aims to help these engineers gain a better understanding of the conceptual bridge design process, helping them to realise why certain designs are chosen.

The project has been underway for some five years, during which time a complete system has been developed. During the first year of the project, background research was carried out and necessary industrial links established. The second year was devoted to knowledge elicitation and the analysis of the acquired knowledge. Using this information, the first prototype was built in approximately three months. Once this had been built, another three months were spent doing the preliminary evaluation and verification. Following this, the developed system was released to other companies so that it could undergo a more rigorous evaluation process. This evaluation enabled the system's applicability to be accurately assessed as well as ensuring that the knowledge base was complete and correct.

This system is still undergoing extensive evaluation and consequent adaptation in industry. Currently, the system is being rebuilt to include a new interface which incorporates some of the comments which have been received during this evaluation and it is hoped that, once this "re-build" has taken place, the system will undergo further development to include sophisticated graphics, which will improve its status in the practical environment.

The system has also instigated further work on bridge aesthetics. During the development of this system, the aesthetics of the bridge were found to be a particularly interesting area of the design process, which was in need of further investigation. Consequently, some new research into this subject has just been started in Cardiff.

Choice of Domain

The domain of conceptual bridge design was originally chosen for a number of reasons. Firstly, there are numerous programs which cover the analytical stage of the bridge design process (i.e. where the suitability of the bridge components is checked and accurately sized, and the overall bridge is dimensioned) and CAD adequately caters for the drawing and detailing stage. However, to date the conceptual stage of bridge design has been largely ignored. This could be because it is an area of bridge design which requires a large amount of practical experience and can therefore only be effectively carried out by senior, practising bridge designers. Little formal training seems to contribute to a bridge designer's development in conceptual design and much of the experience gained is not passed on to younger engineers. Indeed, most young engineers have a very poor understanding of the conceptual design process, which could be attributed to their education or to the nature of their initial industrial experience. It was felt that some way of storing all this expertise and passing it on to younger, less-experienced engineers was required – which made conceptual design a suitable target area for the development of a KBS.

Initially, it was intended that the entire conceptual bridge design domain would be covered. However, at an early stage it became apparent that this domain was extremely large and that to try and create a KBS which would cater for any given situation was unrealistic.

It was decided that it was preferable to stick to a limited but realistic domain rather than be over-ambitious and risk developing an ineffective or, in the worst case, unreliable system. It was recognised that the domain could always be extended at a later stage if the system proved to be successful.

Therefore, in conjunction with our experts, the system domain was limited to road bridges crossing another road. This section of the original domain was chosen as these are the bridges most commonly encountered by the young engineers who were the target user. The considerations which the domain covers are shown in Fig. 8.12 (p. 187).

The domain covers the initial choice of style of bridge (i.e. continuous or simply supported, cable stayed etc.) and the bridge components (that is, the type of deck, the end supports, number of intermediate spans and so on). However, it does not, as yet, cover preliminary member sizing.

Bridge design expertise covers a very broad range of subjects including costing estimates, aesthetics and fundamental structural knowledge. The system incorporates all of these factors, but, as yet, the costing estimates included are not very detailed.

Aims of the System

The primary aim of this project was to prove the viability of applying KBS techniques to engineering design, because at the time, the utility of KBS techniques for design problems was seriously doubted and there were no applications which proved otherwise. Of the few design applications which were available, none were practically oriented nor had any of these applications been tested in industry.

From the outset, the system was developed entirely in conjunction with industrial practitioners as it was the intention of the project to try and develop a system which would be sufficiently reliable, robust and user friendly to be of real use in a bridge design office. Such a system would be of little use to experienced engineers who are confident in their design skills, but to an engineer who is inexperienced in bridge design it would act as a useful tutoring system as well as a design aid. It was never intended that this system would present the user with a final "solution" to a bridge design problem: the system aims to nurse the users through the bridge design process, showing them the considerations involved, while providing them with a viable design. It was always recognised that this design would require further verification before it could be accepted as a final design, but, as the system is aimed at young engineers, it was anticipated that the design would need to go to a senior engineer for checking before going on to the analytical stage.

The system aimed to act as a "pure" expert system (Miles and Moore, 1991) in that it intended to encapsulate and manipulate human expertise in the area as it currently stood. It did not aim to expand or improve the existing expertise.

Sources of Knowledge

Literature on conceptual bridge design is very scarce. Therefore, very little information could be derived from books, papers or from Codes of Practice. We realised that the bulk of the information on conceptual bridge design would have to come from the experts themselves. However, any British Standards and Department of Transport rules which are applicable are incorporated in the system.

As the system was industrially oriented, it was very important to involve a number of bridge design experts in the overall development of the system. However, it was decided to concentrate on using one expert for the bulk of the KBS, as this is the generally recognised approach (Moore and Miles, 1991a). Initially, after a number of experts had been interviewed, a suitable expert was found and he was used for the KE; but he was so frequently unavailable (owing to the nature of his work) that another expert had to be included in the KE to ensure the required progress. Consequently two experts were predominantly used for the KE. This was going against standard practice, but it was a decision which proved to be particularly advantageous for this domain (Moore and Miles, 1991a).

Additional experts were also used to a lesser degree to support the KE. During the evaluation process, two other bridge design experts were introduced. This was an important part of the system development as it helped to ensure that the knowledge base was complete, correct and less biased than would have been the case if only one expert had been used. By introducing new experts at the evaluation stage, the system received a more rigorous review (see Chapter 5 for a more detailed explanation of this).

During the evaluation of the system, a number of other engineers were involved who ranged from potential users (that is, inexperienced engineers) to experienced senior engineers. Their viewpoints greatly enhanced the evaluation process: helping to identify gaps and misleading areas in the knowledge base, hence ensuring the applicability of the system. In total, approximately 15 to 20 of these additional engineers were involved in reviewing the system.

User Interaction

Concentrated user interaction is essential for the development of practically oriented systems. Frequently in the past, users have not been given sufficient emphasis, the focus of attention being the experts and their opinions. With this approach it is unlikely that the system will be acceptable in the working environment, so the potential users must be clearly identified at the start of the project. In our case, the users were chosen to be inexperienced engineers as these were the section of people who would find the proposed system most useful. These users were involved throughout the system development, but most prominently during the evaluation where they were encouraged to influence the development of the system. Their involvement was invaluable and extensive user involvement would be recommended whenever possible.

Style of Knowledge Elicitation

Rapid prototyping was not used for the development of this system as, despite it being recommended (Hayes-Roth *et al.*, 1983) it was found to be an ineffective approach to system building for this domain. Originally, rapid prototyping was attempted as it was the recognised approach. Following the initial interviews, a considerable length of time was spent trying to build the first prototype. It soon became apparent that the size and nature of the subject domain precluded the development of a sensible model using the sparse information which had been acquired, as the information only covered a very small subsection of the entire domain, and many of the connections between the elicited information would need to be estimated. The degree of estimation and guess-work which was required to create a coherent prototype was unacceptable, and it was apparent that a very large number of "preliminary" interviews would be required before a sensible first prototype could be built. This somewhat contradicted the rapid prototyping approach, and by using this approach there was a high probability of producing a weak and ineffective prototype which would risk losing the confidence and co-operation of the expert engineers involved. The preferred approach was to elicit the majority of the expertise before attempting to build the first prototype. Using this approach, there was more chance of developing a coherent and realistic prototype which could be effectively reviewed by the experts involved and which would hopefully increase their confidence in the project as opposed to diminishing it.

Finally, the experts involved in the project, although familiar with computers, were not computer literate. Introducing a computer program very early in the project risked alienating the experts, by making them feel ill at ease with the knowledge elicitation.

This alternative approach to system building proved to be particularly advantageous. The knowledge base which was incorporated in the first prototype was found to be virtually complete, with very few gaps or mistakes being noticed by the reviewing engineers. Using a delayed approach to prototyping enabled the system to be developed in a coherent and logical way, as well as giving the system developers time to analyse the type of knowledge which was involved and hence choose an appropriate form of knowledge representation. It also forced the use of a number of intermediate representation techniques which greatly aided both the clarification of the domain and the eventual encoding of the system.

The decision not to use rapid prototyping as the main approach to knowledge elicitation inevitably affected the overall methodology adopted. As rapid prototyping was not being used, some other way of stimulating and maintaining expert interest and enthusiasm was needed.

We also had to ensure that the experts recognised the project's progress (rapid prototyping automatically gives this impression of progress). To accommodate this, a number of different KE techniques were adopted, some of which acted as feedback procedures, enabling the expert to experience variety as well as progress. The KE techniques used were direct interviews, cards sorts, paper models, pictorial representations and summaries. Techniques which were primarily visually based were deliberately chosen, as design is a task which relies on pictures and sketches, and engineers tend to use pictures to help them understand problems and explain their answers. All of the approaches adopted proved to be effective for the domain in question, but inevitably, with individual experts, some were more effective than others (Moore, 1991).

The interviewing techniques used ranged from unstructured to structured interviews, with the most highly structured interviews being used at the end of the KE period. The effectiveness of this methodology was indicated by the apparent completeness and accuracy of the knowledge base. The completeness of the knowledge base was established during the evaluation, where very few gaps or mistakes in the knowledge base were identified.

During the KE, all the interviews were carried out on a one-to-one basis. This was not exactly by choice: it was difficult enough to get one expert at a time: getting both experts together at once seemed impossible! However, if it could have been arranged, a brainstorming session may have proved valuable. To a certain extent brainstorming occurred during the evaluation of the system, where a number of engineers were invited to assess the system. Their opinions were reviewed using group discussions where various opinions were contested and discussed and new ideas for the development of the system were put forward, which effectively acted as a brainstorming exercise.

Questionnaires were used, but these were largely ineffective as they did not capture the experts' interest or enthusiasm.

Case studies were not adopted during the KE as we felt that the case-specific nature of the knowledge which would be elicited did not warrant the considerable amount of time which would be required to prepare and complete one case study. However, a number of case studies were used during the evaluation of the subsequent prototypes. These were effective for eliciting additional information, particularly exceptions to the rule. Case studies were also useful for identifying the exact limits of the domain.

In total, around 25 hours of interviews were carried out with the experts involved in the KE, over a period of approximately 8 months. This does not include the time spent on transcript creation and analysis or the considerable amount of time spent on encoding and evaluation.

System Development

Once something approaching the maximum amount of feasible information had been acquired by purely interviewing the experts, the next stage was to build the prototype from the elicited information.

The knowledge had to be encoded in such a way that the complete system successfully encapsulated the experts' knowledge as well as efficiently manipulating this knowledge to reach conclusions appropriate to the domain. The encoding of the first prototype took approximately three months. At this time we were under pressure to develop a prototype in a relatively short amount of time as the experts were anxious to see a working system. This obviously affected the way in which the system was built, and a simple approach was deliberately used to ensure that the system was built within the allotted time. Although this approach has proved to be effective in terms of the system's appearance and efficiency, the programming style adopted is not ideal, as the knowledge base is not transparent and the rules are order dependent. We believe that one of the main disadvantages of not using rapid prototyping is that experts can get impatient, so forcing the prototype to be built quickly. However, in our opinion the advantages of not using rapid prototyping far outweigh the disadvantages for this particular domain.

Knowledge Representation – Which Development Tool? This system has been developed using LPA PROLOG in conjunction with a PROLOG-based toolkit: FLEX. There were a number of reasons for this decision. Firstly, when this project was started there was a fairly limited choice of shells available, and those which were available were too limited for the nature of the application. In addition, as this was essentially a research exercise (although it had practical connotations), the exact size and form of the final system was unknown. It had always been the intention to investigate various user interfaces, knowledge representation strategies and reasoning techniques. Therefore, the flexibility afforded by using a language was preferable to adopting the more restrictive format of a shell. PROLOG was chosen as it is a language which is designed for the development of AI applications and it avoided many of the difficulties which would have been encountered if an algorithmic language had been used.

LPA was the version chosen as it is a near standard PROLOG which incorporates the facility for creating "stand-alone" versions of the developed KBS (that is, versions of the system which could be easily ported onto other machines without the need to import PROLOG itself). As we were working directly with industry, we recognised the need for industrial evaluation so this was an important consideration.

Another important factor was that LPA PROLOG was PC based. When this project was started, the only hardware which was readily available in the engineering offices being dealt with was PCs. Also, the majority of engineering offices used a DOS operating environment, which also restricted the versions of PROLOG which could be used.

LPA PROLOG has the added advantage of linking with a toolkit FLEX. This meant that the benefits of a language could be realised while employing the programming advantages of a toolkit.

To a large extent the decision to use PROLOG has proved to be appropriate for the project. However, memory limitations have caused problems during the later stages of the system development. These problems were primarily due to the 640K DOS limit. However, the size and nature of the system (i.e. design is a very wide ranging domain), the software and the fact that we were restricted to a PC also contributed to the problem. These limitations have caused considerable development difficulties as they have not only slowed and affected the development of the system, but they have also caused the system to be unreliable and have, more importantly, prevented the inclusion of graphics in the system. It is recognised that effective graphics are a vital part of any design system. However, the above limitations have prevented them from being included to date. Fortunately, more recently a new version of the software has become available, helping to overcome some of the difficulties which had been encountered. Also, workstations are becoming more accessible within engineering offices, so there is now the possibility of moving the system to workstations. Both of these options would overcome the current difficulties.

Structure of the Knowledge Base
The system is basically modelled on a tree structure (Fig. 8.12), which was created from the smaller tree structures which were used as the paper models during the KE process. This approach was chosen, as during the KE, the paper models proved to be an effective way of representing the elicited knowledge (Figs. 8.13 and 8.14).

The system is split into six sections which directly follow the approach adopted by the bridge design experts questioned. The sectioning helps the user to follow this approach more clearly as well as acting as an indication of their progress through the system.

The six sections are:

Section 1:   Road Geometry.
Section 2:   Topography of the Site and Bridge Width.
Section 3:   Pier Positioning.
Section 4:   Ground Conditions and Bridge Characteristics.
Section 5:   Choice of End Supports and Wing Walls.
Section 6:   Choice of Deck Material and Construction.

The type of questions which are asked in these sections and the conclusions which are reached are detailed later on. From the input, suitable types of foundation, end support and deck are chosen, resulting in a complete conceptual bridge design.

The system currently contains around 75 questions, of which 20 to 30 are asked during an average consultation. The system takes approximately 15 minutes to run.

Which Knowledge Representation Format?
The system relies on the Horn clause predicate style of the PROLOG language, which essentially follows the IF-THEN format of rule-based representation. Although the FLEX toolkit was used, very few of the FLEX predicates were actually employed. The predicates used were those which greatly eased the programming without detracting from the control of the system. For example, an in-built questioning procedure was employed.

Currently, the system contain some 580 clauses, of which two-thirds

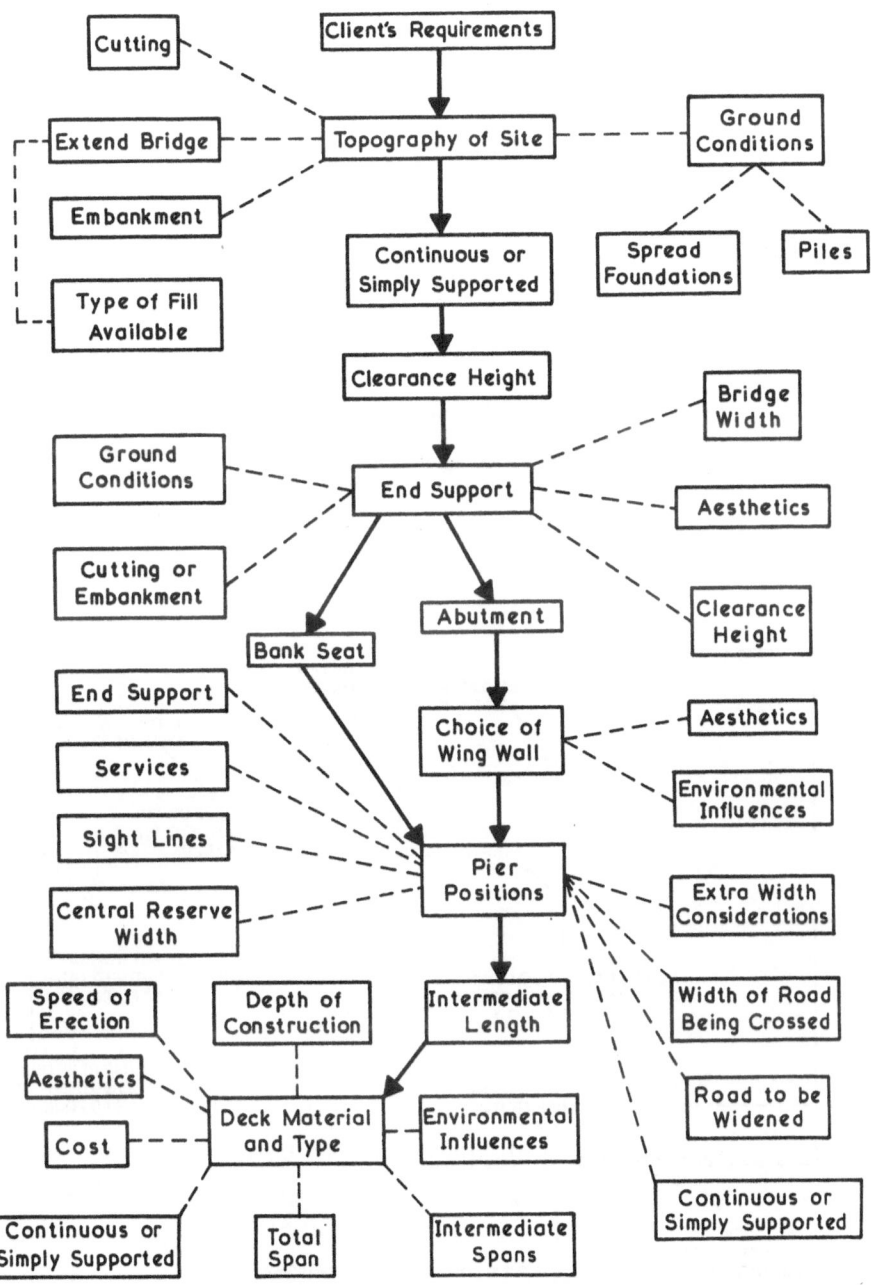

Fig. 8.12 The domain structure for BRIDGE

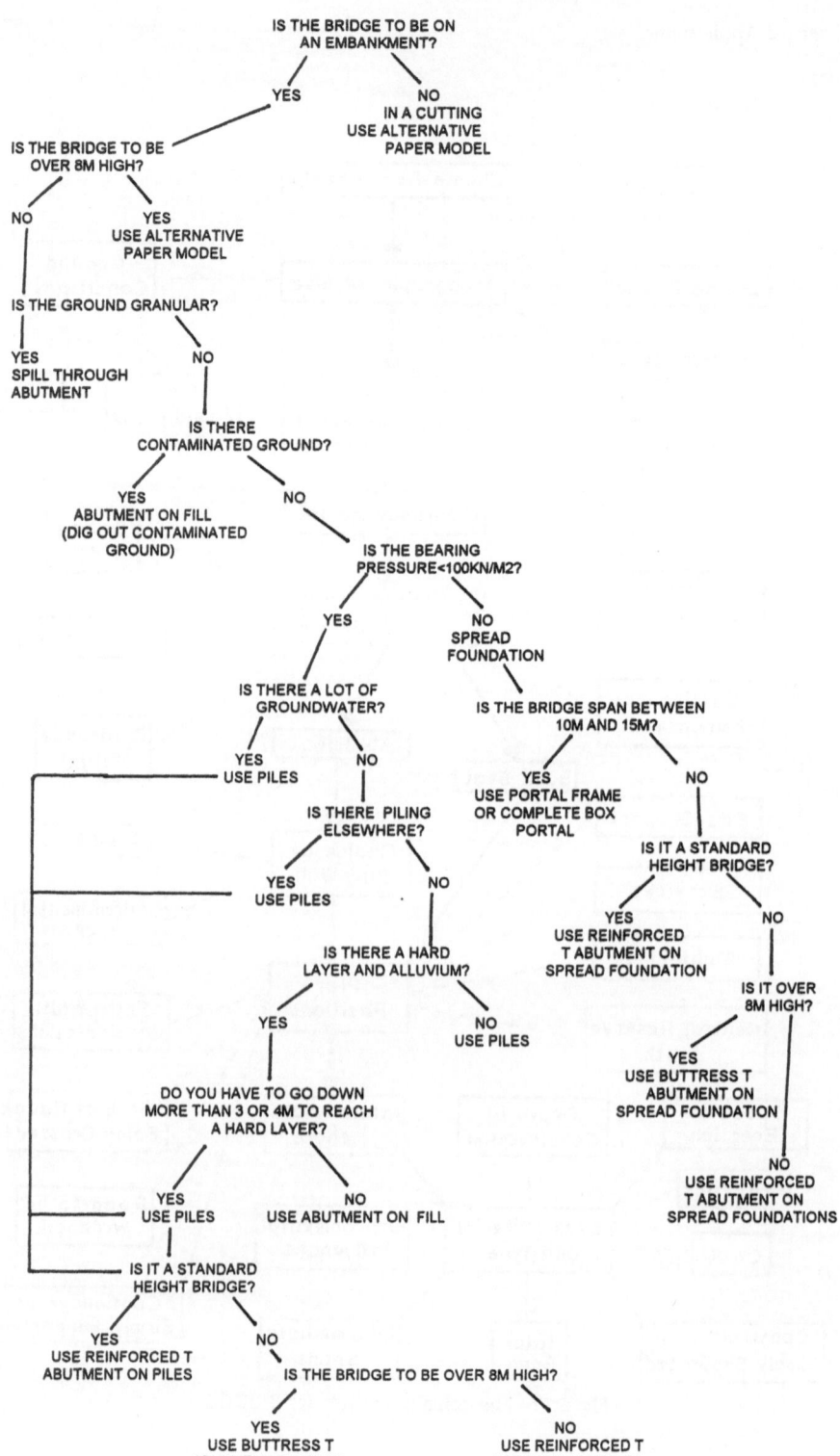

Fig. 8.13 A paper model of the knowledge elicitation used for BRIDGE (end support decisions on embankment)

IN A CUTTING:

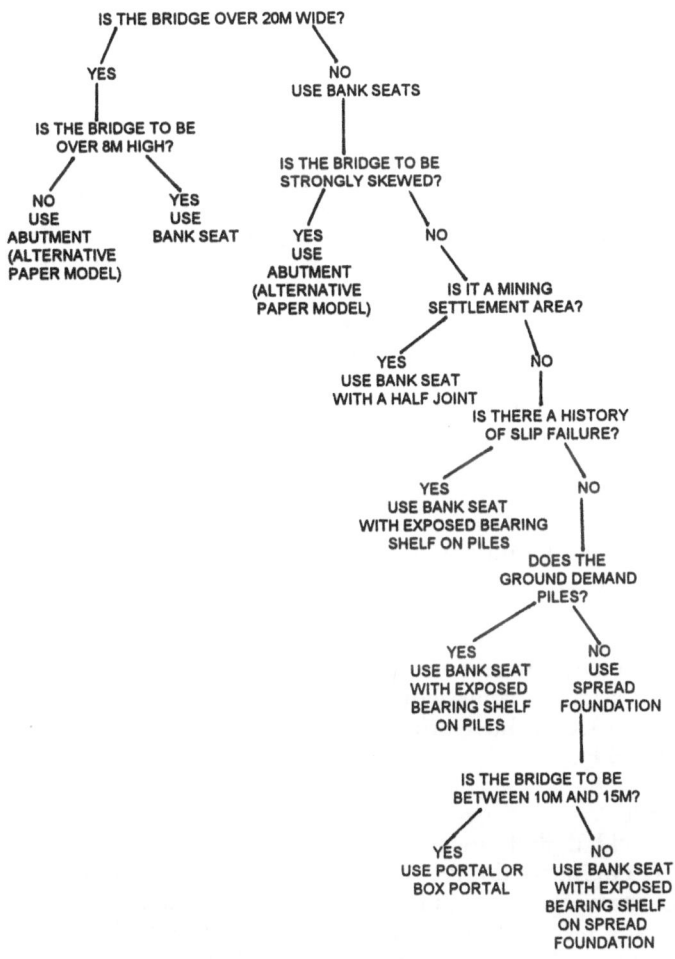

(N.B. FOUNDATION INFORMATION FROM OTHER PAPER MODELS)

Fig. 8.14 A paper model of the knowledge elicitation used for BRIDGE (end support decisions in cutting)

are domain rules. The remainder of the clauses are purpose built to handle the input data and the procedural aspects of running a consultation (for example, for running the help facility).

A typical rule is:

```
searchdeck (L,"to use a box girder"):
        answer(question50,yes),
        member([span,S],S > 30),
        answer(question52,yes).
```

where the answer predicates are inevitably reliant on the user input.

Control of the Knowledge Base

During the KE it was established that expert bridge designers consider the design problem from ground level upwards. As one of the aims of the system was to teach the user about the conceptual design process, then it was important that the system accurately depicted the approach an expert would use. This had to be taken into account when the system was constructed, so that the system asked questions in an appropriate order. This inevitably affected the search and control mechanisms used, as the system had not only to control the search of the knowledge base, but also the display of questions. Therefore, a straightforward backtracking technique was not appropriate. This problem was overcome by using a set of preliminary questions, which are asked at the beginning of every consultation to establish the basic information about the site (in much the same way as a human engineer would). Then, meta-rules are used to control the movement through the knowledge base, to ensure that the questions are asked in an appropriate order while ensuring that the complete knowledge base is searched.

The Rule Base

The rules in the system are order dependent. This is not perfect: ideally the rules would be independent of their position in the knowledge base. However, the size and complexity of the knowledge base precluded this approach. Computer memory problems were encountered using rules which were order dependent: if non-order-dependent rules had been used, obviously the size of the knowledge base would have grown considerably, as more rules would have been needed to cover every eventuality. This would have affected the feasibility of applying the system on an MS-DOS PC.

Uncertainty

It was decided that uncertainty was inappropriate for this domain, as generally engineers do not attach a level of uncertainty to the conclusions they are drawing: a design is either feasible or it is not, and the only other contributing factors which make one design preferable to another tend to be subjective, which it is virtually impossible to quantify.

In fact, in our experience, during the KE an evaluation measurement of uncertainty never arose as an important factor, which is an interesting point, tending to indicate that in many domains its importance is overemphasised.

System Features

User Interface. As the system was intended to be a tutoring aid for inexperienced engineers as well as a design tool, an effective user interface was of paramount importance.

The interface was originally designed with such users in mind and it was intended to display the information in a way which seemed to suit the domain and the type of users. During the evaluation it became apparent that the user interface was of great importance and the standard of interface greatly affected the users' overall opinion of the system. It was

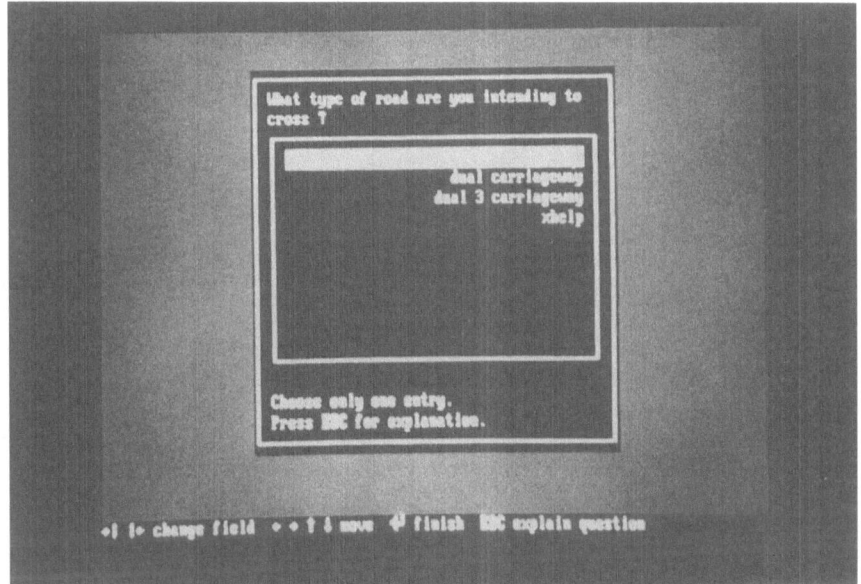

Fig. 8.15 The original interface of BRIDGE: question menu

found that approximately 50% to 70% of all the comments made concerned the interface. It was realised that a more in-depth analysis of effective interfaces for engineering users was required.

A project was started in conjunction with the Psychology Department in the University of Wales at Cardiff which aimed to take a closer look at the style of interface which would be most successful. Interviews with engineering users were carried out so that the interfaces with which they were already familiar could be assessed (i.e. the software which was already available in their offices). Various interface styles were also tested by user trials. According to the information derived, a new style interface was developed (Philbey *et al.*, 1991). The old interface and the new interface are now described. Examples of the old interface are shown in Figures 8.15, 8.16 and 8.17.

The interface was initially based on a question and answer format which was basically menu driven (Fig. 8.15) with only one menu being shown on the screen at any time. Free input (Fig. 8.16) was minimised whenever possible in order to provide an easy-to-use program and to help minimise errors. The help facility was accessed by selecting help from the question menu and then choosing the appropriate option from a pull down help menu (Fig. 8.17). During the assessment of the two interfaces, this pull down menu proved to be very popular with the users, as it is very easy to use and relies on selection using the arrow keys, as opposed to typing in an appropriate number or letter, as is the case with the new version of the interface.

It can be seen that the new style of interface is very different (Fig. 8.18). With this interface, the question is shown at the top of the screen and the explanation is also shown on the screen at the same time. There is also a

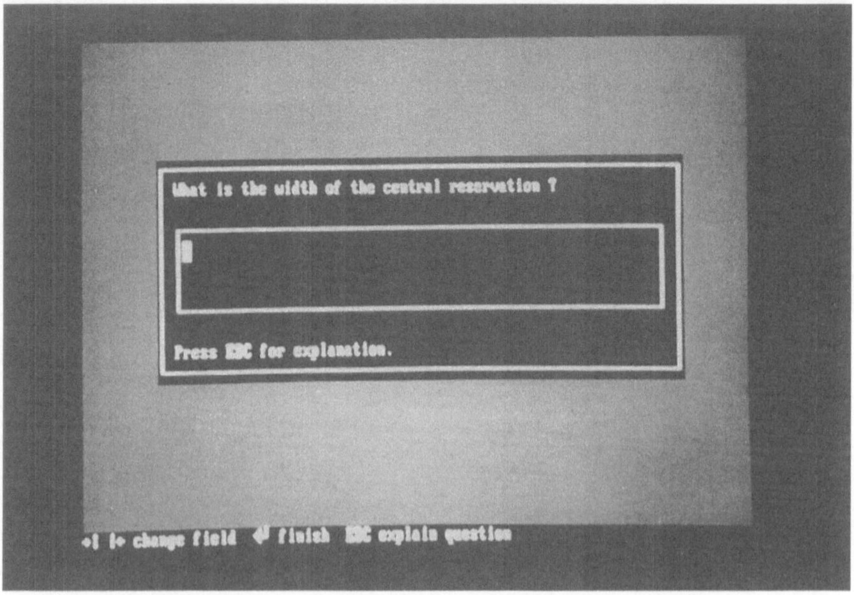

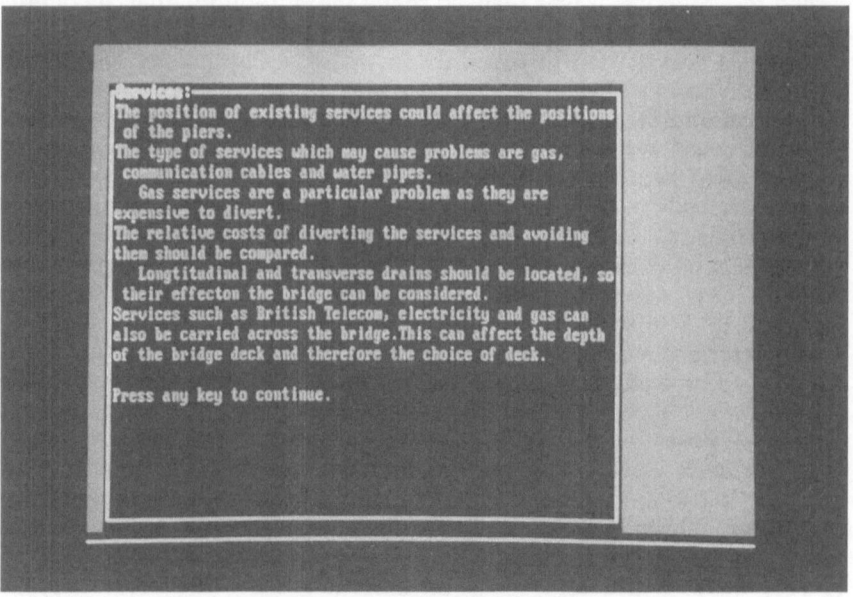

Fig. 8.16 The original interface of BRIDGE: free input questions

Fig. 8.17 The original interface of BRIDGE: help screens

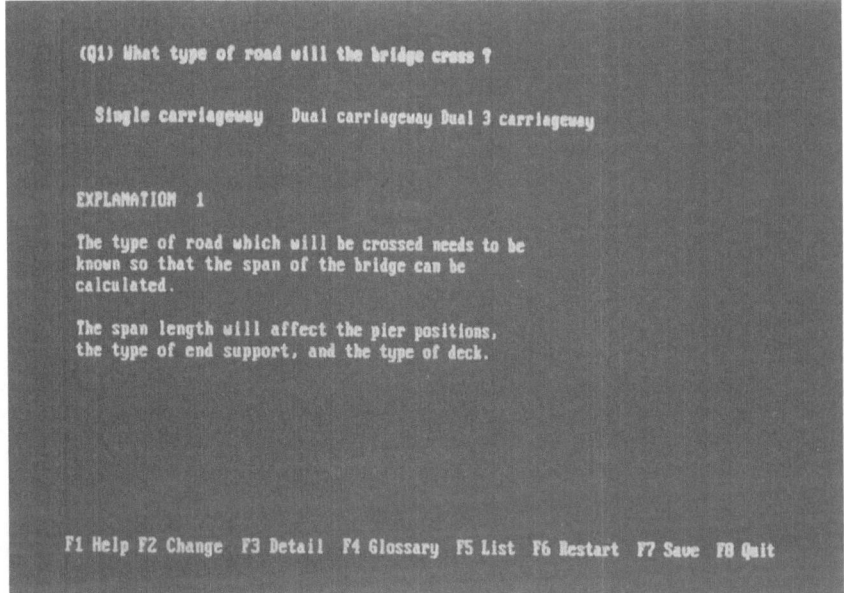

Fig. 8.18 The new interface for BRIDGE

menu on the screen which lists the options which are available (i.e. the help facilities). When one of these is selected, this option also appears as a box on the same screen. A different set of colours has been used for this interface: the impact of these colours on the users' attitude to the system has also been assessed. This interface has the advantage that everything which is available to the user is shown on the screen at the same time. Trials are still underway, but as yet it has been shown that the users prefer the new version of the interface. The system is now going to be rebuilt to incorporate this new style interface. This rebuild will also provide the opportunity to improve the programming style of the knowledge base and to incorporate some additional features.

It must be noted that, as yet, neither user interface incorporates graphics. This is an obvious deficiency in the system, and inevitably graphics are required if the system is to be effective. The incorporation of high-level graphics has to date been precluded by the memory restrictions mentioned earlier. However, it is hoped that when the system is rebuilt to incorporate the new interface, and by changing the hardware from PCs to workstations, effective graphics will be possible.

The Help Facility. The system incorporates a help facility (Fig. 8.19) which can be called at any time throughout the consultation. This facility includes a glossary, which provides a dictionary of terms contained within the system (Figs. 8.20 and 8.21). This is an important feature when dealing with inexperienced engineers, as some phrases or key words can be misleading or confusing. This facility was noted by the reviewers of the system to be particularly useful.

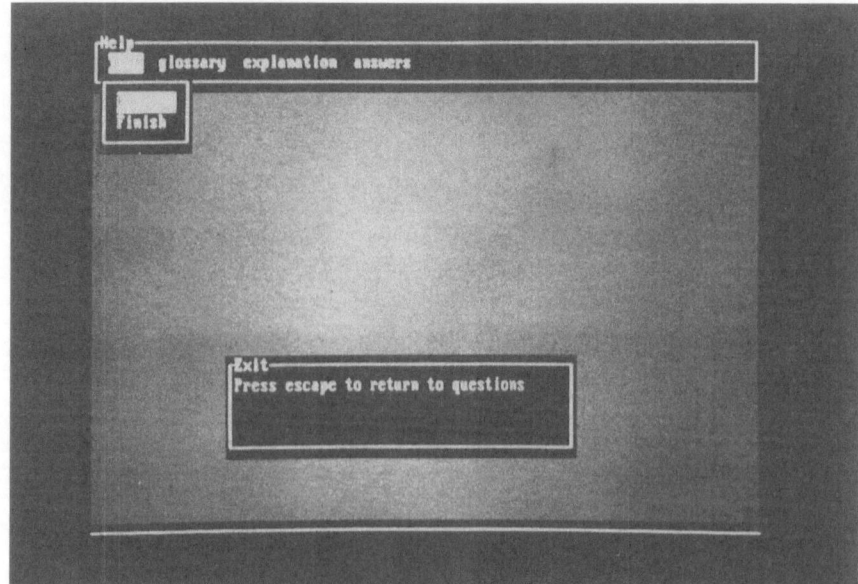

Fig. 8.19 The help menu for BRIDGE

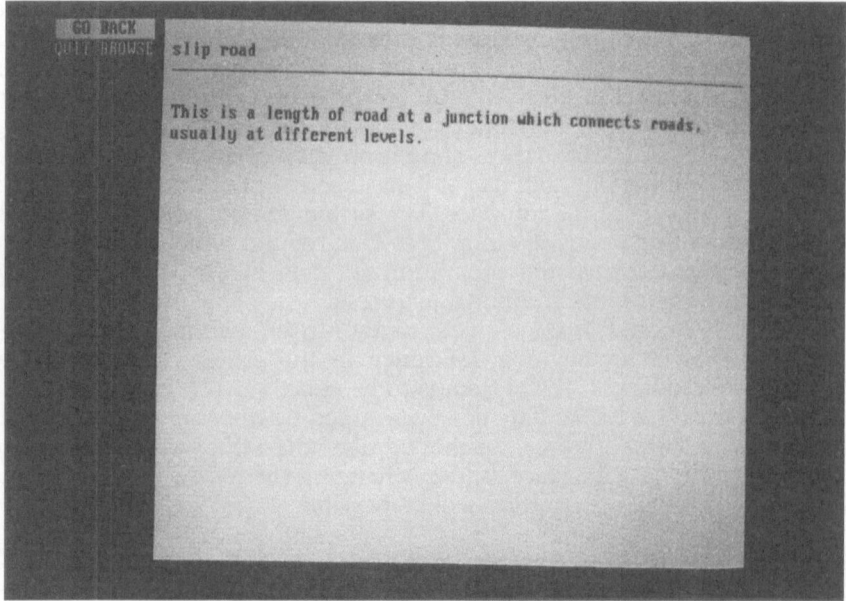

Fig. 8.20 The glossary for BRIDGE: browse format

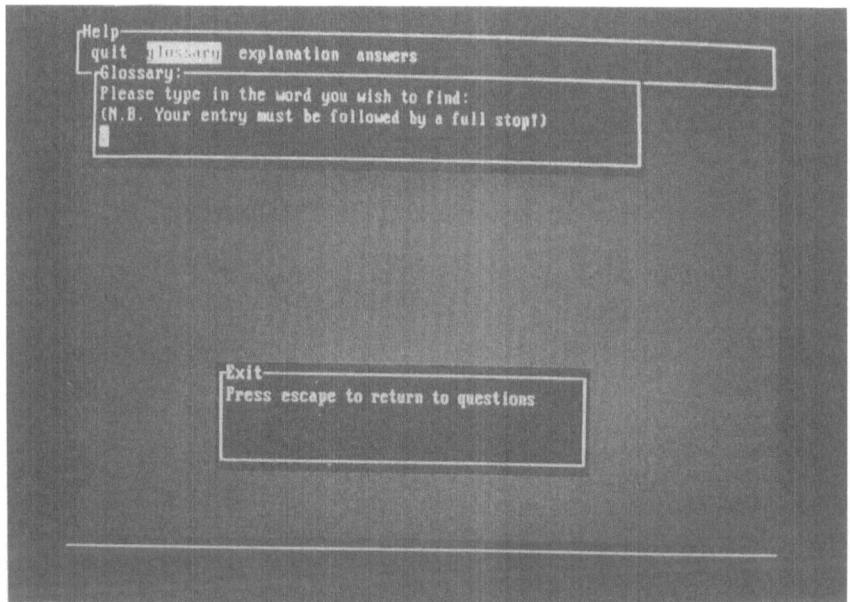

Fig. 8.21 The glossary for BRIDGE: individual word search format

There inevitably has to be a quit facility which allows the user to exit the system altogether or to restart a consultation. There is also a facility which enables the user to review the answers which he or she has already given. This doubles as a memory jogger, showing the route which the system is taking and helping the user to identify the relevance of particular questions.

An explanation facility is seen to be one of the most important features of an KBS (Hayes-Roth *et al.*, 1983; Kidd, 1985; Waterman *et al.*, 1986; Greenwell, 1987). We felt that, because of the inexperience of the target users, the explanation facility in this system should provide more than the standard trace of questions asked and rules fired. Therefore the system incorporates 90 separate text files which describe certain decisions which are taken by the system. During the consultation, help files which are appropriate to the route being taken are stored as a list. When the user requests an explanation, this list is called and the files are printed out in order to provide a continuous string of text (the files are designed to link together so that they form a coherent explanation). This explanation describes the decisions, highlighting the information on which these decisions are based. This provides a full and easily comprehensible explanation, which is important if the system is to fulfil a teaching role effectively.

The explanation facility also incorporates an explanation of the current question which explains the question in more detail, giving a suitable range of answers and explaining its context. This again helps the user to gain a clearer picture of the design methodology being followed (Fig. 8.19).

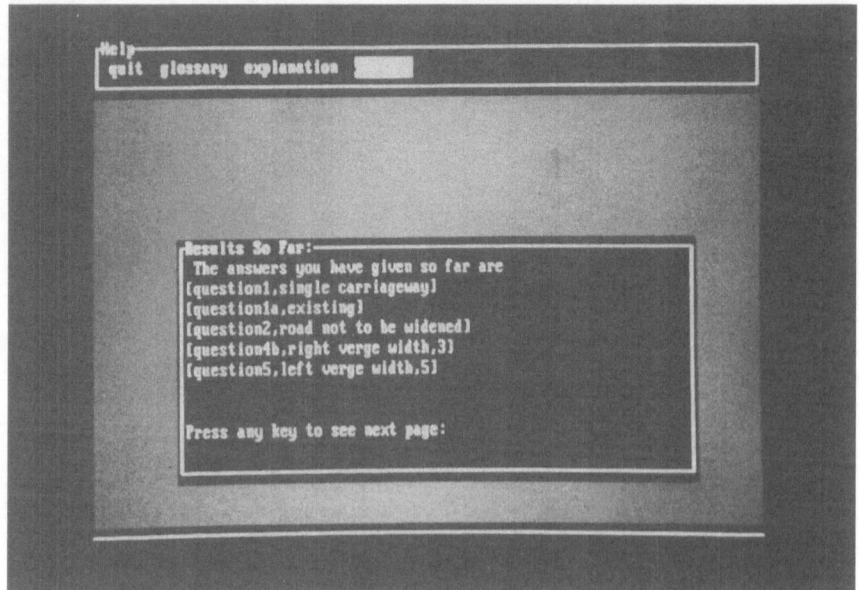

Fig. 8.22   Example list of previous answers for BRIDGE

During the evaluation of the system, many of the reviewers requested a facility which would enable them to go back and change previous answers. This facility has now been incorporated within the system. The previous question can be called and the answer changed (this is in case the user has made a mistake or has misinterpreted the question). Alternatively, any previous question can be called at any point during the consultation. The system will recall this question and ask it again. The user inputs the new answer and the system will return to the point at which the user left off. In some cases, a change of answer will introduce the need to ask additional questions, or will change the decisions which have already been taken. In these cases, only the new questions are asked and the new conclusions are shown. If there are no alterations to the original route, no information is shown and the consultation continues uninterrupted.

At the end of the consultation, the user is shown a list of all the answers which he or she has given, and is then given the opportunity to change these answers. This enables the user to see the effect of small changes to the input on the overall solution. This is a very important feature of a tutoring system, and one which has been well received by the system reviewers (Fig. 8.22).

An Example of a System Consultation

As has already been stated, the system is split into six sections. These sections are dealt with in turn by the system, and conclusions are drawn from each section. Owing to the nature of the system, all of the information which is input by the user is stored so that it can be used in subsequent sections; so in some sections very few or even no questions will need to be asked to reach a conclusion, as the required information is already available.

The system begins with an introductory screen explaining what it is trying to do. This is followed by Section 1: Road Geometry. This section consists of a set of preliminary questions which are asked at the beginning of every consultation. These establish the basic site description so the width which needs to be crossed can be decided. Questions are asked about the type of road being crossed, the slip roads, hard shoulders and verge widths.

Section 2 covers Topography of the Site and Bridge Width. This section establishes whether the road is in a cutting or whether an embankment is needed. It also establishes the dimensions of the cutting and the type of fill which is available.

Section 3 decides approximate pier positioning. It asks questions concerning site lines, road curvature and the width of the central reservation. Inevitably, the verge widths (which have already been input) are taken into account and feasible pier positions are decided. They are not given at this stage, however, as the intermediate span distances are influenced by the type of end support which is not decided until later in the consultation.

Section 4 covers Ground Conditions and Bridge Characteristics. This is currently the weakest aspect of the system. Ground conditions are a very complex area of civil engineering, In fact, even experienced bridge designers tend to consult specialist geotechnical engineers on this aspect of the bridge design. It was initially intended to build a separate system for dealing with ground conditions, but this has not as yet been possible. Currently, the system merely asks a few simple questions about the site and decides whether a spread footing is appropriate or whether piles are needed. It is recognised that this is inadequate. However, it reflects the basic judgements which would be made by bridge designers at this conceptual stage. This section also covers other aspects of the bridge, such as the skew, curvature and bridge width. These are asked at this time as they affect a number of the bridge components.

Section 5 deals with the choice of End Supports which can be used. Frequently, no questions will be asked in this section, as most of the relevant information has already been input. The system will suggest a type of end support (i.e. a type of abutment or bank seat) which is compatible with the bridge and the site characteristics that have already been decided. If abutments are used the spanning distance of the bridge is reduced, so once the end supports have been decided, the intermediate span distance can also be selected by the system by following the rules in the knowledge base. These are also given at this stage. Approximately 10 types of end supports are available as options within the system. This choice is dictated by such things as contaminated ground, bridge width and aesthetics.

Section 6 deals with the Choice of Deck Material and Construction. Again, depending on the previous input, very few questions may be asked at this stage. However, if more information is needed then additional questions will be asked. The system contains approximately 40 choices of deck, and the available materials include steel, composite, concrete and weathering steel. Again, this choice is dependent on many criteria such as aesthetics, economics, the surrounding environment and spanning distance.

Once the deck has been decided, the overall design is summarised for the user. Obviously, graphics would make this summary much clearer. Work is currently underway to develop suitable graphics for the presentation of the final design. However, at present, only a summary of the numerical data is given. This is followed by the user being given the opportunity to change some answers. This allows the user to "tweak" the input to assess the effect of small changes on the overall design. The user is also given the opportunity to start again.

The system deals with the information in the same way as an expert designer would: from ground level upwards, which makes it an effective teaching system. This is unusual for a KBS: usually the search is much less constrained, and consequently the questions would appear in a more random order. Although this would have been efficient it would also have been confusing for the user, and would not have accurately depicted the design process.

Evaluation

The evaluation which this system has undergone has largely been described in Chapter 7. It is the only engineering design system that we know of which has undergone an in-depth process of evaluation, verification and validation, and as such its practical applicability has been markedly improved. The evaluation process also enabled the developers to learn much about the way in which the applicability of the system works, how to improve it and how the system should develop in the future.

The system evaluation showed that the knowledge base was virtually complete, as no major gaps or mistakes were found by the reviewers. This confirms that the KE was effective.

The preliminary evaluation stage involved the system being checked by the experts who were originally involved in the KE. Once they were satisfied that the system was correct and fairly represented a section of their expertise, the system was released for evaluation by a wider range of reviewers. These included experts who were new to the system, potential users and a number of engineers of varying levels of experience. By using a variety of reviewers, a wider range of opinions could be obtained, helping to ensure a more objective system.

The system was placed in a number of bridge design offices, where the reviewers were allowed to use the system at their leisure. A diary was provided in which they could note their comments. Approximately once a month, the reviewers from each company would meet with the developers to discuss their findings. These meetings proved to be as productive as the diaries in eliciting information and ascertaining the reviewers' opinions of the system.

The evaluation showed that the users were strongly influenced by the user interface, as has been discussed earlier. It also showed that their enthusiasm towards the evaluation noticeably improved once they realised that their opinions were of importance and that their suggestions were being incorporated in the system. Originally the less experienced reviewers were wary of giving their opinions. However, after a few discussion sessions, they were keen to contribute more to the system's development (Moore and Miles, 1991c).

Case studies were used during the evaluation process. To date, the system has been tested on some 50 to 60 case studies, during which time it has exhibited an 85% success rate. A success is taken to be when the system produces a design which is the same as the original, or one which is agreed to be viable by the experts involved. A failure is when the system could not or did not produce a correct solution.

The system is still undergoing evaluation in conjunction with a review of the new user interface. It is hoped that this review process will continue once the system has been rebuilt, to enable its practicality to increase.

Future Developments

Although at present the system only caters for small to medium span road bridges crossing a road, it has been developed with expansion in mind. It is intended that the system will be extended to include loading configurations, preliminary member sizing and more detailed costings, helping to make it more useful. It is also hoped that the system's domain will be extended to cover railway bridges and footbridges.

In the more immediate future, the system will be rebuilt to incorporate the improved user interface. This rebuild will enable the structure of the knowledge base to be improved so that the rules are less order dependent and more transparent. Inevitably, graphics must be included in the system in the near future. Now that the memory problems have been overcome, this should not be a problem.

It is hoped that eventually the system will be developed to interact with other computer systems, such as intelligent databases and CAD systems, to provide a more global design advice system. With the improved availability of workstations within the engineering design offices, by moving the system over to operating on workstations, this interaction will be made much easier.

Case Study 3: A Knowledge-Based System for the Strategic Planning of the Disposal of Sewage Sludge

Introduction

This system is somewhat different from the above examples which are firmly rooted in structural design. The title "strategic planning" at first implies that this is something other than design, but in fact strategic planning is conceptual design; it is just that this particular part of the water industry uses a somewhat different vocabulary from that used by most designers.

The work on the system started in late 1989. The work is partially sponsored by the British Water Research Centre (WRC) and they initially had a requirement for a system to help with the catchment (basin) planning of sewage effluent discharges. As will be explained below, this aim was altered so that the work has resulted in a system for sewage sludge disposal.

The system is written in C and runs on any PC which uses the MS-DOS operating system, although some parts of the software (more specifically the analysis) can be a little slow on some of the less powerful machines. Our co-worker on the development of this system is John Hooper, and credit for much of the detailed work goes to him.

As yet, the system has not been tried on users. The reasons for this omission are because of the problems with establishing the domain, which are discussed below. However, the development work on the system is proceeding and it is hoped to start user trials in the near future.

Initial Work

As stated above, when we were given the brief to develop this system it was for the development of a KBS for the strategic planning of the Best Possible Environmental Option (BPEO) for the disposal of sewage effluent (i.e. the liquid part of the treated sewage as opposed to the solids). The work started in a typical fashion for the development of a KBS by locating suitable experts (in this case two) and going through the standard procedures of knowledge elicitation. As suggested by Moore and Miles (1991a), the "standard" AI technique of rapid prototyping was not followed, but instead an attempt was made to collect a substantial body of knowledge before any programming was attempted. Meetings with experts were difficult to arrange because of their commitments elsewhere but gradually over a period a number of interviews took place.

As system developers, we had been led to believe that the knowledge required to define a BPEO strategy for effluent disposal existed. The experts who participated in the knowledge elicitation believed that they used such an approach and obviously our sponsors thought that designers followed such a strategy. However, our knowledge elicitation revealed that the BPEO is not considered and indeed the body of knowledge required to define the BPEO does not exist. Given that those who were sponsoring the project and the participating experts were genuine in their beliefs, the above finding is very interesting.

What actually the designers currently do is find the cheapest scheme (in financial terms) which also satisfies the relevant legal constraints. Most of these refer to the environment and so some account is taken of environmental factors, but inevitably when one is aiming for a least cost solution then the environmental restraints tend to be satisfied at a level which is very close to the allowable minimum. In such a situation there is no need for a KBS because the decisions are made by calculating the net present value of each option that satisfies the legal constraints and picking the cheapest. There might be a minor area for the use of KBS technology in helping to set up the options, but such a system would be relatively trivial and certainly not worth the effort of a major research project.

The early history of this work is presented as an object lesson regarding some of the problems that can occur during the development of KBS. As we discovered, designers do not always follow the procedures/strategies that they believe they have pursued (for possible reasons, see Chapters 2 and 4) and therefore in knowledge elicitation it is prudent at an early stage to analyse the knowledge gained to check the validity of the proposed domain for the establishment of a knowledge base and subsequently a KBS (Miles *et al.*, 1990).

Setting up the Domain

Having identified that the original domain did not contain a sufficiently mature body of knowledge to define a KBS, we were faced with abandoning the project or finding a related area. The obvious place to look was in the disposal of the solid part of the sewage (well, relatively solid, sludge can contain as much as 98% moisture by weight). We managed to find two experts, Dr Colin Powelsland, who at the time was an employee of our sponsors, and Dr Neil Harkness.

Obviously our first job was to check that the domain of sludge disposal was suitable for setting up a KBS and that the expertise to establish a BPEO existed. This process caused us some concern, but when we analysed the transcripts from the early knowledge elicitation interviews with our experts, they seemed to be full of heuristics; and also the influence of cost in sludge treatment and disposal is less than for effluent treatment and disposal, because the costs between the various options tend to be the same order of magnitude. This has enabled a body of expertise on BPEO for sludge disposal to be established.

It has to be admitted that our confidence had been badly shaken by the early difficulties and so we had some further misgivings, but as we got deeper into the knowledge elicitation our concerns disappeared.

System Objectives

The broad aim of the system was, as stated above, to produce a KBS which would advise designers who wished to ascertain the BPEO for sludge disposal. The system was aimed at the part of the process when an overall strategy is being determined rather than the detailed stage and, as already discussed this is analogous to conceptual design in other domains.

It next becomes necessary to look at the possible users of such a system and at what level they would wish to interact. Two major types of user were identified:
1. The designer, that is, the person who will do most of the work.
2. The manager and his peers or superiors. The manager is the desig-

ner's boss, who needs to be able quickly to assimilate the various items of data, form a mental picture of how they relate and then convey that information to other decision makers within the organisation.

Thus the system had to be created so that it could deal with the designer at a relatively detailed level and yet output the data on which its decisions are based, and the decisions themselves, in a manner which is capable of interpretation by people who cannot afford the time to study the details.

Also, it was decided that it would be useful if the "managers" could, after the designer has input the basic information, use the system to play "what if" games, to test out different ideas and check the sensitivity of the problem being studied to various changes in parameters. Therefore the system needed to allow relatively easy access to the knowledge for a given design problem.

Hardware/Software

At an early stage, the decision was taken to develop the system on a PC. The advantage of this is that the software is readily portable. Thus, when either the sponsors wished to see what progress has been made or the experts were asked to review critically the work to date, the software could either be demonstrated on a lap-top or installed in a suitable machine. Workstations are not as common as PCs, and then there is the problem of the different operating systems and internal configurations that are found, which further limit portability. In the engineering industry, certainly in Britain, PCs running under MS-DOS are found in virtually every office. The disadvantage of undertaking all the work on PCs occurs during the development process when, on occasions, the speed of a workstation would be useful.

A number of options were examined for the software. Initially PROLOG was considered, but it was felt that after our experience with the bridge system, trying to develop a KBS in another language would give us a useful insight. PASCAL was then considered, but finally rejected in favour of C. The choice of the latter in preference to PASCAL was somewhat arbitrary, but we have as yet no cause to regret the decision.

In the KBS literature, the transparency of the knowledge base (i.e. the fact that it can be read and understood easily, especially by those who are not familiar with the programming language used) and the fact that this allows users to access and change the knowledge base are generally quoted as being major advantages of knowledge base programs. Our previous experience has, however, shown that it is not a good idea to allow unfettered access to knowledge bases. System developers spend a great deal of time collating and checking the information that subsequently becomes the knowledge base. They then spend even more time making sure that the final system behaves as it should do. Subsequently to allow users to alter and possibly corrupt the knowledge base is not a good idea (there are possible ways around this problem which are discussed in the following chapter).

In the light of the above chain of thought, no attempt was made in the sludge system to make the knowledge base especially transparent. To an outside user, the system looks just like any other C program.

Knowledge Elicitation

The problems described above with the initial knowledge elicitation exercise dictated that for this system, unlike our previous work, where we had elicited the bulk of the knowledge before programming, a technique more akin to rapid prototyping was used.

Initial interviews with our experts were conducted using standard "shut up and listen" techniques to allow the experts to set the domain. After only three interviews programming started.

The domain was such that it could be easily compartmentalised into the following categories:

1. Physical properties of the sludge.
2. Disposal options, these being: using agricultural land, landfill, incineration, land reclamation and composting.
3. Sewage works data, including sludge treatment plant.
4. The environmental impact assessment.

After the three interviews, a paper model of the overall domain and its structure was constructed. The experts were asked to check and amend the model as appropriate.

Following this, the knowledge elicitation took the form of further in-depth interviews asking for specific information about given areas of the process. Only one area was worked on at a time, so initially work started on the sludge's physical properties, the sewage works data and then the disposal to agriculture. The latter proved to be a large and complex topic and has occupied most of the time to date. Intertwined with all this work was the environmental impact assessment which is so intrinsic to the whole process.

Once a section of the system had reached a reasonable standard, it was demonstrated to the experts and their comments were elicited. So the approach used formed a sort of halfway house between rapid prototyping and eliciting the bulk of the knowledge before starting programming.

The method of working suited the domain because it could be compartmentalised fairly easily. There was a stage fairly late on in the initial stages of software development when demonstration of the system to one of the experts elicited a large amount of new information, which called for quite substantial alterations to the program. This was very time-consuming (about 2 months work). It is interesting to speculate on whether or not a more thorough initial approach to knowledge elicitation might have avoided the problem, but without a control one can only guess.

Knowledge Base

The range of knowledge that the system incorporates is quite wide. There is a great deal of numerical data on such things as sludge moisture contents and concentrations of various chemicals, the latter relating to the sludge itself and also to the various disposal routes. Thus a substantial degree of the processing undertaken by the system is algorithmic.

Also, in addition to the knowledge gained from the experts, an appreciable volume of knowledge has been acquired from the literature. Some of this knowledge comes from Codes of Practice and legislative requirements and the rest has been extracted from research papers and text books. For example, the assessment of the volume of sludge that can be placed on agricultural land is based on a formula which is taken from the work of the Soil Survey of Great Britain.

Our experts have then input a substantial amount of heuristic knowledge, which is used to judge such things as the final moisture content of a sludge after it has been through a belt press or to what level a sludge has to be dewatered before it becomes suitable for incineration (about 26% moisture content by weight is about the maximum level at which self-sustaining combustion will occur). Other heuristics have been elicited regarding the assessment of environmental impacts.

One of the great disciplines of trying to express a method of solving a problem using computer code is that everything has to be fully defined. One area which this work has shown was not very well defined, but which is one of those areas that humans cope with very well, was the comparison of the various environmental impacts between different options. The heuristic assessment procedure is discussed in the following section, but it is useful to consider one particular aspect of that work here.

Take an example where there are ten viable options for disposing of the sludge. If, say, for traffic movements the difference between the highest and lowest options is substantial but a few of the options are very similar, then the problem of comparison becomes very difficult. A simple ranking of the options will fail to express the fact that there is a great deal of similarity between some of them. So it is necessary to devise a scoring system which reflects this. As yet, work on this aspect of the system has not been finalised. The procedures used in the heuristic assessment follow the standard techniques of environmental impact assessment, but the advantage of the system is that it enables sensitivity analyses to be undertaken very rapidly, with several analyses per minute being possible.

The system does not have a knowledge base in the conventional sense that one associates with AI systems. Instead the knowledge is embedded within the code of the program. This makes the system more difficult to understand for the casual user who picks up a source code listing, but the use of C has made it possible to build a fairly substantial system within the DOS 640K limit and the software has proved to be far easier to debug than, say, a PROLOG program.

System Features

The system is split into a number of linked C programs. Sitting above the whole system is a file manager which enables the user either to access an existing data file or to create a new file.

If a new datafile is to be created, then default values are available for all the unknown parameters. As there is a substantial amount of input data, the number of times the default facility can be accessed is large and therefore the reliability of the final answer can be undesirably low. At the moment the system is incomplete, but it is intended to weight the final answers for a given option by taking into account the number of times that the analysis of the option has relied on default data.

Beneath the file manager sits the system proper. This has three major modes of operation:

1. Input (primarily of data but also of system parameters).
2. Analysis.
3. Output.

The Input. The input section, as its name suggests, controls the input of data to the system. There are four categories of input, these being the physical properties of the sludge itself, the characteristics of the potential outlets (e.g. areas of farmland, the heavy metal concentration levels on the farmland etc.) and the parameters relating to the sewage treatment works and the processing of the sludge therein. There is one further part of the input section which is not related to the sludge itself but to the balance between the various sectors of the assessment of the options.

When choosing a sludge disposal option from among a number of other options there are three categories which can be considered:

1. Environmental impacts.
2. Security of operations.
3. Costs.

With costs it is obviously possible to add up all the numbers and finish with one overall price. This is the beauty of anything which can be expressed as a number and preferably in common units. Hence the power of statisticians over such vulnerable people as politicians!

With the other two categories, things are not that simple. For example, the environmental impact is made up of a number of subcategories:

1. Odour.
2. Visual.
3. Traffic movements (caused by road tankers moving sludge).
4. Land pollution (e.g. heavy metal build up in soils).
5. Air pollution (e.g. from traffic movements generated by the sludge disposal).
6. Ground water pollution.
7. Surface water pollution.
8. Public acceptance.

There are also eight sub-categories under the heading of security of operations.

The problem arises that it is necessary to compare options on a sensible and logical basis. This comparison has to include costs, environmental impact and security of operations, and all the subcategories. As discussed above, with costs it is easy because they will all reduce to a single number but with, for example, environmental impacts how does one rank odour

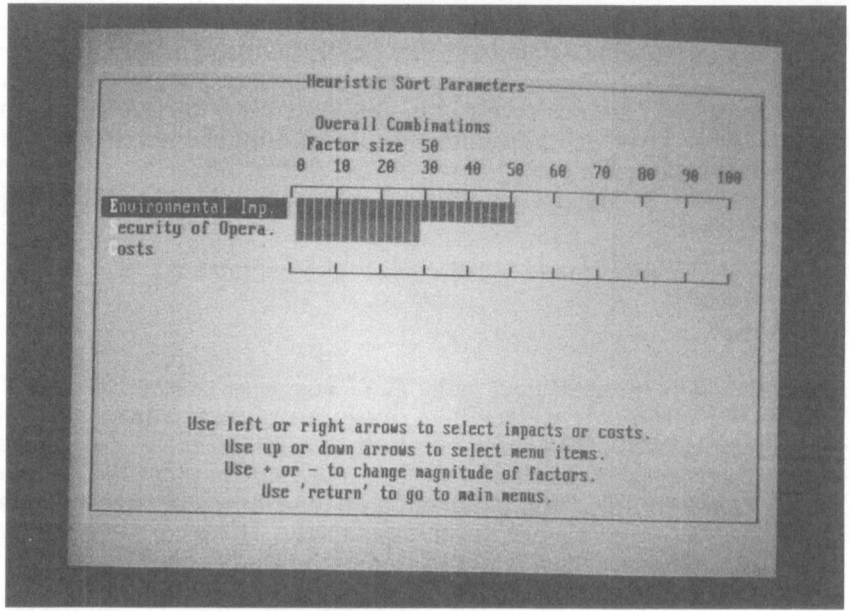

Fig. 8.23 Heuristic assessment, mechanism for weighting main components: sludge disposal system

and visual impact? Furthermore, how does one then rank these in comparison with costs? Such problems are common in environmental impact analyses.

The approach that has been adopted in this work is based on the idea of a graphic equaliser of the sort that is commonly found in sound systems. The technique has been given the name of heuristic assessment. Each of the sub-categories can be weighted within a given range, as is shown in Fig. 8.23. The subcategories for environmental impacts, costs and security of operations can be seen. The technique allows each scheme to be given a single score, although what the units of that score are is impossible to say. When one can reduce everything to a number, then it is possible to rank options in order of priority. Also, if the ranking appears to be wrong, then one can use the system to investigate why and alter the factors, but in so doing one has consciously to upgrade certain factors and downgrade others, so the implications of various value judgements are made explicit.

The use of this approach has been very interesting. To date, sludge disposal planners have never been asked to quantify or justify their decisions, and so they would make statements of the form "Odour impact is important here because of the proximity of housing" but they would then never rationally weight that impact in comparison with other factors. Thus in this case the development of a KBS has forced the planners to think hard about the implications of their decisions and also to question and expand their expertise. To date, a definite answer on the weighting of the various factors is not available, and probably never will be. Each user of the system will want to see the sensitivity of the options to the weighting of the various factors.

The Analysis.   When considering the disposal of sludge to agricultural land, various factors need to be optimised. Typically only about 10% of farms within an economic travelling distance from a sewage works will accept sludge. What is therefore necessary is to identify these farms, assess what size of sludge tanker can be used for a given farm (e.g. if the access is poor then only small tankers can be used – this is more expensive than large tankers) and then check on the suitability of the land for sludge disposal. Of course, if the land quality is not good enough, then that farm is removed from the analysis. Obviously none of the above process is undertaken by the sludge system. This is something for the designer to sort out.

At this stage then, one is left with a number of farms (typically between 10 and 50) at varying distances from the sewage works that are able to accept a given size of sludge tanker. The system now checks that the land is suitable for sludge disposal (e.g. not too close to houses, not too steeply sloping, not too wet and having suitably low pollutant concentrations). Having made these checks, ideally one would like to find the optimum combination of farms for disposal. Historically, experts have used their judgement to arrive at a figure, but, using the idea of heuristic replacement, as described in Chapter 2, the sludge system has a built-in facility for arriving at an optimal answer. This requires minimising the cost of the sludge disposal (taking into account different tanker sizes, number of journeys, etc.) and ensuring that the capacity of the chosen farms to accept sludge is equal to or greater than the sludge output of the sewage works.

The optimisation uses a Genetic Algorithm (GA) (Goldberg, 1989), the basis of which is described by Hooper *et al.* (1992). Typically for a 20 farm problem, the GA on a 386 PC will find the optimum solution within 5 seconds. Obviously, for larger problems the solution may take a couple of minutes but procedure is very fast. This then is an excellent example of how when developing a KBS it is possible to identify an area where human processing power is insufficient to solve the problem accurately, and yet a computer because of its ability to process data at high speed can easily provide an accurate answer.

The Output.   As stated above, one of the aims of the system is that it should provide information on its decision-making processes at a level which can be quickly assimilated by people who are not involved with the details of the work but may be involved with the decision making regarding the final choice of sludge disposal options. We have chosen to call these people managers.

Some parts of the system such as the heuristic sort, are by their nature readily accessible to all users, but the area where the most work has taken place is that of the output from the system.

The output facility is split into five sub-categories:

1.   Sludge.
2.   Outlets.

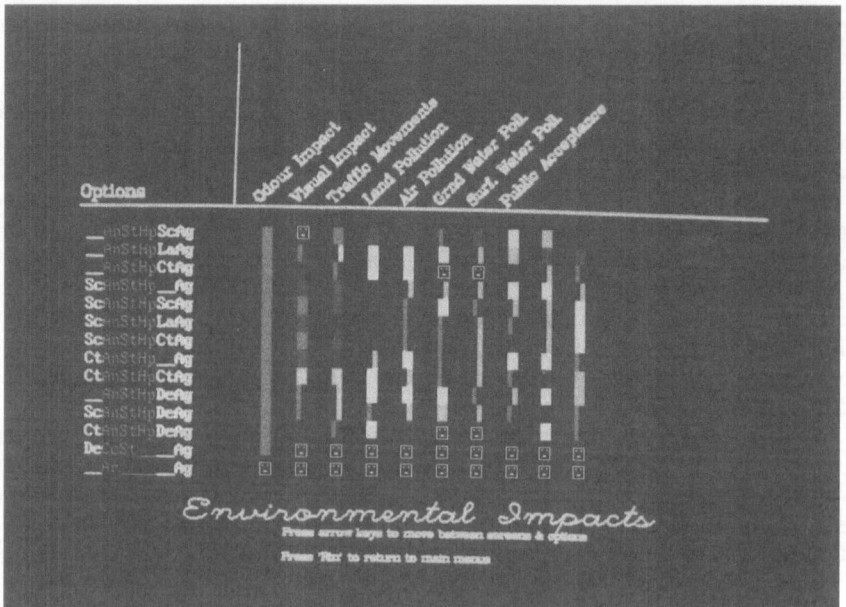

Fig. 8.24 Options summary, chart for environmental impacts: sludge disposal system

3. Treatment.
4. Evaluation.
5. Final results.

In all of these sections the data inputs to the system regarding the given study are summarised in fairly conventional forms using graphs, tables of statistics, pie charts and bar charts. There is one unusual feature which so far as is known is unique in computer systems (although common in some consumer magazines) and that appears in the Evaluation sub-category where, for the main features of each possible option, colour-coded charts are presented so that at a glance it is possible to see which features within a given option are good or bad. The charts cover the same areas as the heuristic assessment: environmental impact, security of operations and costs. The cost chart obviously just gives the actual figures, but the other two charts use the colour coding.

For example, Fig. 8.24 shows the chart for the environmental impacts. The symbols on the left-hand side refer to the various combinations of treatment that are to be used in that particular option. For the visible symbols the explanation of their meaning is as follows:

Ij – Raw liquid sludge is injected into soil.
Li – The sludge is treated with lime.
St – The sludge is stored.
An – Anaerobic digestion.
Hp – Combined heat and power from the sludge.
De – De-watering of the sludge.

To the right of these symbols are eight columns. Each column refers to an environmental impact and for each option shown on the left-hand side there is a colour-coded rating of the environmental impacts. The colours range from green for excellent (shown as pale grey in Fig. 8.24) to blue for very poor (shown as dark grey in Fig. 8.24). The layout of the chart allows the user to see quickly what the strengths and weaknesses of a given option are. Thus the three evaluation charts for costs, environmental impacts and security of operations allow someone who does not want to get involved in the detail to understand the system's ranking of the options.

An Example

As with the other systems, it is useful to include a "run through" a typical example to show how the system works. The amount of information that is asked for and consequently the amount of output produced by the system is too great to be fully covered, and so only the salient points will be described.

On starting up the system, the initial menu is for the file manager which allows the user to retrieve a particular datafile or to input new data. The next menu is the start of the system proper. The user is offered a number of options. It can be seen that this is a different approach from that used by the offshore design and the bridge design systems, where the first screens were asking the user for information. With the sludge system, because it has been designed to be utilised by different types of user at various levels, the software allows the user to choose what operation is to be undertaken. With the two preceding systems a more directed approach was used, where the user interaction was a question and answer session with the software controlling what the system did next. This raises the interesting question as to which type of interface is the best, but unfortunately we have no information as yet on which to base a judgement.

The menu shown in Fig. 8.25 is the main menu of the system and, as shown, gives the user five options. Obviously the End command takes the user out of the system. The Analysis command runs the procedures which for each option undertake the environmental impact assessment, check for compliance with legal restraints, produce the output data, calculate the heuristic assessment scores and activate the Genetic Algorithm (GA). The GA, as explained above, currently only optimises the choice of farms for the agricultural option, but if the system is extended it is hoped to use this facility to optimise other features. The analysis has to be accessed each time the parameters for a particular scheme are changed. The Files command allows the user to access the file manager.

The remaining two commands are Input and Output and, as one would guess, the Input is for inputting data to the system and the Output is to allow the user to look at the results of a given project. Thus the combination of the file manager and the menu will, for instance, allow someone to use the system to load the data for a given scheme and just review the system's recommendations as to what the impacts and costs of the available options will be.

Fig. 8.25   Basic system menu: sludge disposal system

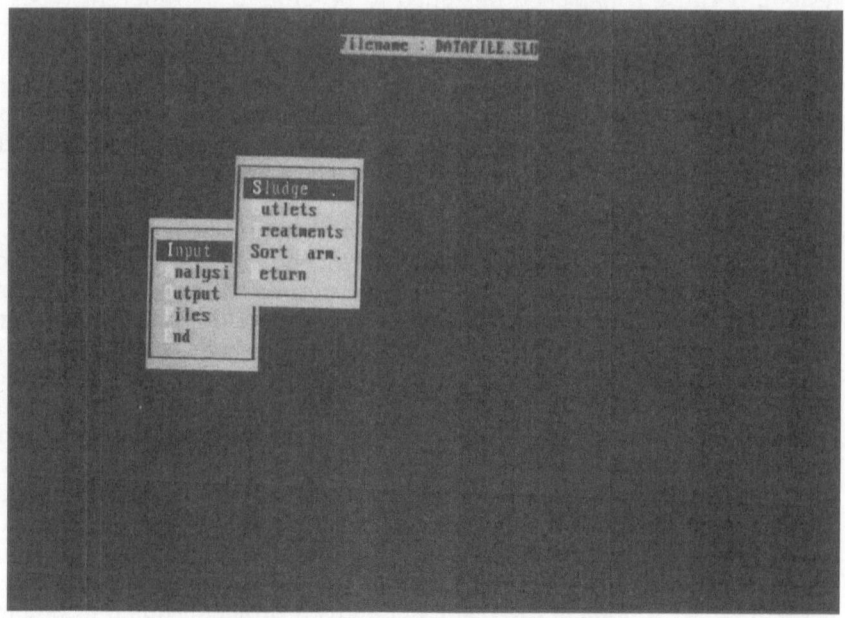

Fig. 8.26   Input menu: sludge disposal system

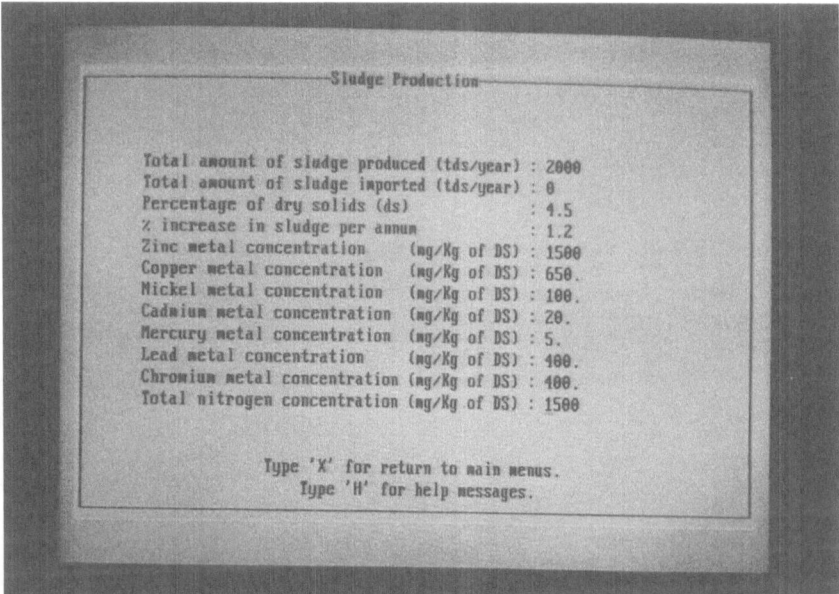

Fig. 8.27 Typical data input menu: sludge disposal system

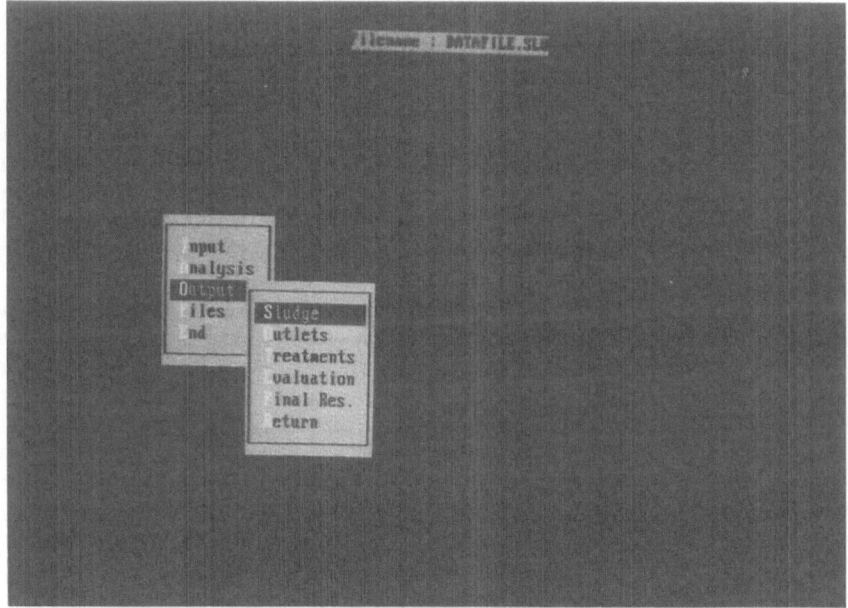

Fig. 8.28 Output menu: sludge disposal system

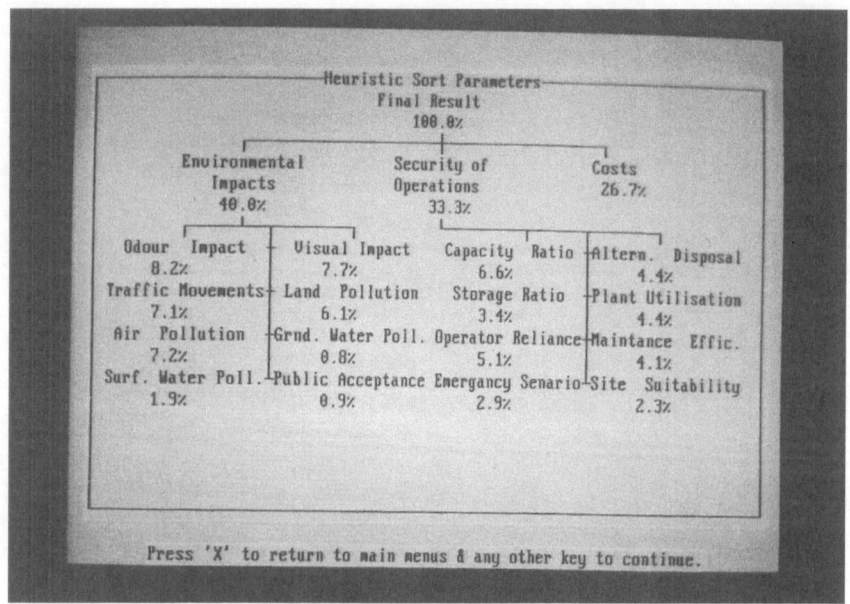

Fig. 8.29 Components of the heuristic assessment: sludge disposal system

If the user chooses to Input information then a further menu of a similar format appears (Fig. 8.26). The user can choose to input data for any of these items. For example, if Sludge is chosen then on the screen appears a series of menus of which Fig. 8.27 is typical. The style of input is chosen so that it is quick, and if the user wishes to use a default value then these are available. This type of interface is far removed from what one would normally associate with a KBS and especially with conceptual design, but the particular style of interface suits the types of data which are associated with the domain. Many of the data are numerical and the remaining inputs are mostly yes/no type answers which allow the system to be designed so that all the input is via menus of the same form.

Choosing the Output option from the main menu brings up another menu of the same style (Fig. 8.28). Each option within this menu gives the user access to data regarding the particular problem which is being worked on. Some of the items on the menu have been covered (for example, the style of the evaluation screens has been discussed above). The other output screens mostly consist of either charts of statistics, graphs, pie charts and histograms. The Final Results item produces data regarding how the heuristic assessment has worked. An example is given in Fig. 8.29, which shows all the items that contribute to the final score given to the options and the organisation of the heuristic assessment.

From the above, it can be seen that the system is very different from a traditional KBS and there is little apparent intelligence in so far as the system does not select what information it needs to solve the problem. This is a feature of the domain. It is necessary to have all the data available before one can begin to make choices. Obviously, depending on the level of definition of the design process, not all the data may have been collected and hence the system provides defaults.

Future Developments

The original sponsors of the work are unable to continue owing to lack of resources, but further money has been made available from other sources and so the system development is assured. The work will continue on the environmental impact assessment, the development of the other disposal outlets and user trials will start in the near future. It is not anticipated that the basic form of the structure will change greatly, but undoubtedly the user trials will have a some impact.

Conclusions

This chapter has given a brief outline of the evolution of design KBS and has shown the way in which three systems which we have developed have evolved and consequently differed from one another. These three detailed case studies exhibit how our opinions of the relevance and future of KBS in design have changed as we have gained more experience and have worked closer with industry, enabling us to identify its needs better. The final chapter discusses these opinions on the future of design KBS in more detail.

Chapter 9

Future Developments

Objectives

This chapter aims to:

- Summarise the current state of KBS, identifying the problems and benefits with current approaches.
- Indicate where the future of practical KBS lies, as we see it.

Introduction

The preceding chapters have concentrated on factual information and, where possible, the discussion has been based on our own experience, as we know these methods and ideas have been tried and tested. However, in this chapter we will indulge ourselves in speculation, although we emphasise that the ideas put forward are based on considered thought rather than mere whim.

Something which is an actuality and not speculation is that, in terms of applying KBS technology to conceptual design, mankind is still close to the bottom of the learning curve. Although in the last decade there has been much research in this area, an agreed strategy as to the best way to proceed has yet to emerge. This is not unusual for a technology which is in its infancy. There seems to be a need to learn about techniques, test them, learn from the associated problems, and then generate and test these new ideas before a recognised approach can be established. Although advances in computing technology may have led to ever faster machines, when it comes to creativity we still have to work at a pace which is dictated by the human brain.

This chapter therefore describes the features which we feel the design systems of the future will or at least should possess. We will look at current developments and give our judgements (for what these are worth) on what will succeed or fail. These views are personal and will inevitably cause some dissent.

The Overall Form of Design Systems

In this section the discussion centres on the complete design process as opposed to concentrating on conceptual design or KBS. This is necessary to achieve an overall appreciation of computer systems and their use in the design process before focusing on KBS.

Data Exchange

Eastman (1981) realised that rather than just providing tools to ease analysis, computers could provide assistance on a much wider scale. Over the last decade, with the increasing implementation of computing and related techniques, the feasibility of this widespread implementation has increased and indeed it can be envisaged that in the future it will become uneconomic not to undertake the design process from first concept through to manufacture on a computer.

Currently, a variety of forms of representation, including mental models, paper representations and computer-based techniques, are used throughout the design process. Using different representations creates a lot of work, and each transition between one form of representation to another introduces possible sources of misunderstanding and consequent errors.

For example, take the creation of a typical building. The forms of input to the design process are as follows:

1. The client's requirements; that is, the purpose and cost of the building. This is generally in text form on paper.
2. Legal constraints and requirements. These can be imposed local agreements or national design codes and other standards. These will be in text form on paper.
3. Geometrical information. These are survey data giving the spatial layout of the site and surrounding features. These may be in paper form either as unprocessed readings or as drawings. Increasingly it is becoming possible to obtain all these data in digital form, and in future it can be anticipated that all land survey data will be in machine-readable form.
4. Geotechnical information. These vary from borehole logs taken by site operatives through to data from laboratory tests. The data are presented in a variety of forms on paper, although work is underway to develop suitable KBS forms of representation (Toll et al., 1991).

A typical design process will start in an architect's office where initial concepts and ideas are sketched out, and the design is then consolidated to a point where it can be passed on (usually in paper form) to structural engineers. The latter have to interpret the architect's drawings and provide suitable foundations and support structures. The building services engineer then has to fit in all the ducting, pipes and cables, and organise the necessary plant before the design can be considered to be viable.

At each stage of transition between these groups of designers there is an increasing possibility of misinterpretation or simple mistakes (both of which can also occur at other times but the probability of their occurrence is less). If the design was largely computer based then it would be possible for many of the inputs to the project to be in digital form.

If this were the case, it would become both desirable and economic for the architect to undertake the entire design, from concept to final drawings by computer. The design data could then be passed to the structural and services engineers in machine-readable form.

In order to extract the full benefit from the computerisation of this process, an additional feature is needed: this being a product model (or models). Product models are described in detail below, but can be broadly conceived of as a means of passing information from one computer system to another, about such things as, for example, what each part of a CAD drawing represents. This enables the receiving computer system to understand and retain these meanings within its own internal representations. Therefore, in the above example, such information would enable the architect to designate spaces so that if any of the engineers then tries to violate the allocated space, the computer system would be capable of remembering the architect's definition and preventing this violation from occurring, hence reducing the number of possible mistakes which could be incurred.

Small is Beautiful?

A number of large scale research projects are underway to create comprehensive design environments which incorporate CAD, various KBS and analysis capabilities. The main advantage of these sizeable systems is that they are (or at least they should be!) compatible, because all the various components of the system come from the same source. However, to date the success record of these ambitious schemes in terms of producing a useful product has been poor. A more worrying disadvantage of this approach is that if a single design system of this type manages to secure a large proportion of the market, then it will effectively become an industrial standard. This introduces the risk of a partial monopoly which would be financially disadvantageous for purchasers and which also could restrict the introduction of new ideas. If one software supplier has a monopoly, then why should it spend more than absolutely necessary on research and development?

The alternative is for a design practice to purchase a number of separate software systems and somehow to link them together. This leads immediately back to the concept of a product model, which would not only work at the level of information being passed between different design disciplines or practices but also as something which facilitates the transfer of information between different design systems. The current problem in most areas of design activity is that the development of product models is still in its infancy, making it difficult for design practices to choose and buy appropriate systems which can be successfully linked.

The best way forward would seem to be the development of a method of readily linking systems. Certainly, small stand-alone software systems are relatively inefficient to use because of the work required to input and extract the information. It would be preferable if the system could obtain the data it needs through direct links to other software. Given this degree of flexibility, design practices could then custom-build systems to suit their own special needs and financial resources.

Product Models

Product models have been mentioned as being an essential factor if the maximum benefit is to be obtained from computerising the design process. In this section we will explore the concept of product models in more detail. Much disagreement still exists about what a product model is: there are almost as many definitions of a product model as there are researchers working in the area. Finger and Dixon (1989b) define them as "models that combine representations of geometry, semantic knowledge and engineering models". A more succinct definition is that a product model should uniquely define or describe a product. However, the degree of information about the product that should be contained within the model is the subject of some debate. Should the product model just contain "sufficient" information or should it be as comprehensive as possible, on the basis that it is difficult to predict what others will need to know in the future?

In effect, product models are a means of giving computer systems some sort of understanding of the information that they are processing. When a designer draws a component, he/she is conscious of whether it is just a preliminary sketch to see how things might fit together or whether it is a final working drawing which is backed up by detailed analysis. The designer also possesses background knowledge about the utility of each of the items on the drawing: for instance, he or she recognises whether it is a screw, a wire carrying an electric current or a flexible material to help absorb vibrations. If we are to build complex design systems, then these systems need to be able to pass such data on to other systems.

So far, we have implied that product models are means of transferring information between different software systems. This is just one of their possible uses. They can also form the internal representation structure that is used within a software system so that the system can understand and reason about the design. Also, they can be used to provide a database during the lifetime of the product. For example, take a building: if the owners of the building had a complete product model of the building then this would greatly facilitate the maintenance of the structure, since if any alterations were required the original design criteria could be accessed. It is recognised, however, that designers may be reluctant to pass too much information on to third parties because of the dangers of litigation and loss of intellectual property.

The main work on product models is taking place under the ISO/STEP initiative. Apart from a few areas where money has been injected by commercial interests, the progress of the work is slow. This lack of progress is largely because of funding problems, but it is also due to the immensity of the task and the amount of research and development work that is necessary. However, it must be recognised that without adequate product models, the full benefits of computer technology in design offices will not be realised.

KBS for Conceptual Design

Having looked at design systems in general in the previous sections, this section focuses specifically on KBS—initially looking at general principles and then concentrating on specific areas.

Prescriptive or Reactive?

Current KBS technology can be traced back to the work of Feigenbaum and others at Stanford University on early expert systems such as DE-NDRAL and MYCIN (Shortliffe, 1976; Lindsay *et al.*, 1981). Expert systems, as their name suggests, were developed to simulate a consultation with an expert, and they were initially developed for diagnostic areas as opposed to design domains. They therefore tend to ask the user for information and then prescribe a solution. While this type of behaviour is typical of such things as a medical consultation, it is not typical of design.

Such an approach may be acceptable for a system which is only used occasionally to obtain advice about a particular area. For example, in structural design, one can envisage the system of McCarthy and Nouas (1991) for unusual types of steel connectors being of great use in this way. The reason that such an occasional consultation would be acceptable is that this is exactly how one uses experts: they are not consulted for everyday advice but on an infrequent basis.

However, our experience of testing KBS in design offices suggests that although the expert system approach is acceptable to those with little or no previous experience and who therefore need a lot of help, other designers find such systems restrictive. More distinctly, the experienced users find this style of system boring to use because it takes control, doing all the thinking for the user and hence reducing the designer to merely a supplier of information.

So how do we see the design KBS of the future? Our ideas on heuristic substitution have already been covered in Chapter 2 and this will be an important factor in the future of KBS, but we believe this will only be part of the story. For design KBS to be successful, it will be necessary for them to behave in a manner which fits in with the design process. Designers are innovative and creative people who are used to controlling the design process, indeed this can be a very stimulating and enjoyable activity. They generate their own ideas which are either based on past designs or their own experience and intelligence. By doing this, a large number of skills are utilised and they can embrace knowledge from many domains, including that all important source of judgement – common sense.

To fit in with this style of working, we believe that the design KBS of the future will act as more of an observer than an expert, hence allowing the user maximum control over the design process. The designer will undertake the design at a workstation or equivalent, sketching out ideas on the screen. By using some form of product model, the design KBS will be able to understand these ideas on the screen and it will therefore be able to observe and comprehend what the designer is doing. If the KBS has nothing to contribute, that is, if the design is both correct and acceptable, then the system will remain passive in the background. However, if the knowledge base can suggest a viable alternative, or if a mistake is detected, then the designer will be prompted by the system. The designer will then have the option of accepting the system's advice or of carrying on with his/her current line of thought. Thus the type of system that is being suggested is more reactive than prescriptive. This type of system not only maintains the designer's creative skills, but also enhances them by ensuring that more feasible options are considered, and increases the chances of detecting errors. Such a system should also be capable of providing assistance to the designer by remembering design decisions. This would help to overcome cognitive overload and also act as a quality assurance mechanism. It would, in addition, also assist the development of a defined design procedure.

As yet, with the exception of Drach et al. (1992), the above ideas on design KBS do not appear to be widely supported. A possible reason for this is that our work has involved a greater amount of testing of design KBS by practising designers than is the case with many researchers who have tended to concentrate on the development of advanced computer techniques. Our work has concentrated more on the needs of our industrial collaborators than on perfecting computer techniques. The interests of the two research areas are inevitably different and consequently the findings differ.

What Form of Knowledge Base?

One of the supposed objectives of KBS technology is that the software allows the knowledge base to be written in a form which is sufficiently close to everyday spoken and written language to enable it to be easily extended and altered by the users of the system. This aim however, has always remained on the wish list of desirable features as opposed to being fully implemented. Knowledge bases for any non-trivial domains tend to be large and complex, and to establish a knowledge base takes a significant effort. To validate and correct the knowledge is then also a major task. The thought of allowing users access to such a carefully formulated piece of work is rather alarming, as it would be very easy to insert information which was in conflict with other parts of the knowledge base forming an incoherent knowledge store. Furthermore, there is the risk that unwittingly the knowledge base might become corrupted by this new information, so that the system will start to give incorrect solutions. This was recognised as a potential problem very early in the development of KBS (Shortliffe, 1976). Although there are techniques in existence for conflict resolution and error detection, generally these problems are difficult to overcome and KBS developers are justifiably reluctant to allow unfettered access to their creations. There is the additional problem that if alterations are allowed to the knowledge base, then who becomes responsible for its performance? Once a considerable number of changes had been made, it would be difficult to assess the applicability and accuracy of the system without a consistent review process, which involved the original developer.

From the point of the designer however, easy access to a knowledge base is a very desirable feature. Design practices earn their living by using their specialised knowledge. They are naturally reluctant to pass this on to third parties such as knowledge engineers, and yet if truly competent design KBS are to be developed then the system developers need access to this knowledge. There are two possible solutions to this problem: either design companies employ knowledge engineers to establish and maintain a knowledge base which is unique to that company or the designers are allowed access to the knowledge base to alter and amend it as they see fit. The latter option obviously clashes with the arguments in the previous paragraph, and yet the former would be very expensive and not overly practical. Case-Based Reasoning offers a partial solution to the above problems but this technique is still very much in its infancy. As it is, we may face the same problems of scale as "conventional" KBS.

An ultimate solution is not yet clear but, given the problems of establishing and maintaining large knowledge bases, one possible route forward would be to attempt to fragment such knowledge bases, as a smaller knowledge base is much more easily understood. However, as yet this technique has not been tried and therefore the effect of such fragmentation on the overall performance of the design system cannot be stated. One can see that with a reactive design system it would be possible to associate various small knowledge bases with daemons which would only be fired if the designer followed a certain route. This method of working would be possible because the KBS would not be controlling the process and therefore it is not necessary for the knowledge bases to be connected.

An alternative is to build a sophisticated input checker which could be used to monitor the changes and additions that are made to the knowledge base to ensure the minimum amount of conflict and corruption occurs. Inevitably the effectiveness of such a checker would be limited and it could not hope to cover every eventuality. However, it would provide a record of all the changes which had been made, making it easier to detect how errors had been introduced.

Uncertainty

One thing that humans do supremely well is to reason with uncertain knowledge, and this is one of the areas in which it is most difficult to get KBS to perform well. The current approach to uncertainty based on a pseudo-statistical approach seems clumsy and difficult both to use and comprehend. Possibly its ultimate failing is that it relies on numerical manipulation, whereas the human reasoning process tends to use an appreciation of the words used and their context. The problems which arise are therefore similar to those faced when using an algorithmic language to represent a non-numeric knowledge base.

Our prediction is that, in the same way as declarative representation schemes have been developed for KBS, the statistical approach will be supplanted by some form of heuristic reasoning which would be more appropriate to uncertainty. Work on various types of heuristic reasoning for design is in progress. Techniques such as these would enable uncertainty to be quantified in a way which is more suited to human reasoning processes.

Combining Advanced Computing Techniques

Finally, it must be recognised that, if the full potential of AI technology is to be realised, then in the future advanced computing techniques, such as KBS, Neural Networks, Case-Based Reasoning and Constraint Logic Programming, should be seen as complementary techniques and not as exclusive approaches. At the moment, there is a great tendency to see each new computing technique which is developed as being segregated from the rest, and this encourages these new techniques to be compared with existing techniques and shown to be better. The majority of approaches and techniques which have been developed to date each exhibit their own advantages and disadvantages. If AI applications in design are to develop and improve, all the available techniques and their associated benefits and pitfalls should be considered, and possible combinations of techniques which venture to overcome the apparent difficulties should be employed wherever possible. Only in this way can the most be made out of current technology and effective systems developed. Combining novel techniques in this way instead of comparing them will in turn encourage the production of systems which are suited to the domain in question, as opposed to proving the viability of specific computational techniques.

Similarly, the advantages of co-operating systems have already been discussed in Chapter 8, and research into this area is already underway (Huang, 1990). Co-operation between existing design computing technology (i.e. CAD and analytical packages), as well as between recently developed AI systems, is necessary if the computerisation of the design process is to succeed, as this is the only way to ensure that all aspects of the design process are adequately catered for without diminishing human innovation and creativity.

Improving the Acceptance of KBS in Industry

One issue which is of paramount importance is the acceptance of KBS by designers and their employers. Inevitably, the introduction of such systems into industry will be relatively slow, in much the same way as CAD was only gradually adopted by practitioners. However, again in a similar way to CAD, their eventual introduction and acceptance will prove to be beneficial. The following sections discuss ways in which this process of acceptance can be accelerated.

Evaluation

The subject of KBS evaluation has already been fully discussed in Chapter 7, where it was stated that effective evaluation is an important facet of KBS if they are to evolve in a sensible, designer-oriented manner. However, practical evaluation is an area of KBS development which has largely been neglected to date. If design KBS are to be fully accepted and utilised in industry in the future, then evaluation must be carefully considered. Only systems which have been developed in close collaboration with industry and evaluated over a reasonable period of time will develop in such a way as to be accepted into their working environment. The evaluation should involve both people and case studies, with particular reference to real designs. In this way useful systems can be developed and new ideas for further development and implementation of KBS identified.

Increased Interaction with Industry

Our work has shown that close collaboration with industry is essential, not only for developing useful systems but also for identifying the direction which KBS development should take in the future if useful systems are to be developed. Increased interaction of this kind is needed to secure the future of KBS in industry and to ensure that the systems which are being developed are applicable and useful in their intended environments. Too many KBS are developed but never actually used, because it is found on their release that they do not fulfil the actual needs which industry has.

Increased User Involvement and Control

The importance of user consideration has already been mentioned in the above sections. Users should be identified and involved in the development of computer systems: otherwise how can their needs be catered for! More importantly, however, the KBS developers of the future must remember that the majority of users are intelligent and innovative people who need to feel involved and have control over the design process which they are carrying out. This can only be achieved by handing the control of the KBS over to the users, making them responsible for the creativity and the innovation while leaving the computer to do what it does best – remembering large amounts of data and identifying mistakes.

Conclusions

As yet, designers have not started to make any significant use of KBS, although there has been a considerable amount of interest from the design community in the work which has been undertaken to date. Techniques have developed to the point where it is possible for KBS to begin to be of real use in conceptual design, although much research is still needed before the full potential of design KBS can be realised.

A great deal of the research effort into novel design systems has concentrated on introducing complexity into existing techniques. This research direction seems intuitively to be misplaced. Until the basics have been established, then making existing techniques more complex is clever but the effort seems to be misguided. Design itself is not yet fully understood, and how KBS technology can best be utilised in a design office is far from being clearly appreciated. Until these fundamentals are established, massive advances in the computing technology will be wasted as they cannot be effectively implemented in poorly understood areas.

References

Adeli, H (Ed) (1988a). Expert Systems in Construction and Structural Engineering. Chapman and Hall, London.

Adeli, H (1988b). AI languages and programming environments. In: Adeli, H (Ed), Expert Systems in Construction and Structural Engineering. Chapman and Hall, London, pp. 23–29.

Adeli, H and Balasubramanyam, KV (1988). Expert Systems for Structural Design. Prentice Hall, New Jersey.

Adelson, B (1988). Modelling software design within a problem-space architecture. In: Design Theory 88, Proceedings of the Second Workshop on Design Theory and Methodology, 56–80.

Adelson, B (1989). Cognitive research: uncovering how designers design; cognitive modelling: explaining and predicting how designers design. Res. Engng. Des., 1, 35–42.

Akin, O (1978). How do architects design? In: Latcombe, JC (Ed), AI and Pattern Recognition in Computer Aided Design. IFIP, North Holland, Amsterdam, 65–103.

Akin, O (1986). Psychology of Architectural Design. Pion Ltd, London, 196pp.

Akin, O, Dave, B and Pithavadianm S (1988). Heuristic generation of layouts (HeGeL): based on a paradigm for problem structuring. In: Gero, JS (Ed), AI in Engineering: Design, Elsevier, pp. 413–444.

Alexander, C (1964). Notes on the Synthesis of Form. Harvard University Press, Cambridge, Massachusetts, 210pp.

Alim, S and Munro, J (1988). PROLOG based expert systems in civil engineering. In: Pham, DT (Ed), Expert Systems in Engineering. IFS Publications, pp. 131–146.

Allwood, RJ (1989). Techniques and Applications of Expert Systems in the Construction Industry. Ellis Horwood, Chichester.

Allwood, RJ, Hinde, C, Negus, B and Stewart, DJ (1985). Evaluation of Expert System Shells for Construction Industry Applications. Loughborough University of Technology.

Allwood, RJ, Shaw, MR, Smith, JL, Stewart, DJ and Trimble, EG (1988). Building dampness: diagnosing the causes. Building Research and Practice, 16 (1), pp. 37–42.

Alty, JL and Coombs, MJ (1984). Expert Systems: Concepts and Examples. NCC, UK.

Alvey, P and Greaves, M (1990). The leukaemia diagnosis project. In: Bramer, M (Ed), Building Expert Systems. Wiley, New York, pp. 15–28.

Anderson, JR (1985). Cognitive Psychology and its Implications. Freeman, New York, 503pp.

Annuba, CJ and Watson, AS (1992). An innovative approach towards designer-oriented CAD systems. The Structural Engineer, 70 (9), 165–169.

Apgar, HE and Daschbach, JM, 1987. Analysis of design through parametric cost estimation techniques. In Eder, WE (Ed), ICED-87, ASME, New York, pp. 759–766.

Archer, LB (1968). The Structure of the Design Process. Royal College of Art, London, 200pp.

Ashley, DB and Wharrey, MB (1985). Prototype Expert System for Subsurface Risk. NSF Grant CEE-8352354.

Ashley, KD and Rissland, EL (1988). Weighting on weighting: a symbolic least commitment approach. Proc. 7th Nat. Conf. on AI, pp. 239–244.

Asimov, M (1962). Introduction to Design, Prentice Hall, Englewood Cliffs, New Jersey, 135pp.

Bachant, J and McDermott, J (1984). R1 revisited: four years in the trenches. AI Magazine, 5 (3), pp. 21–32.

Baffes, P and Wang, L (1988). Mobile transporter path planning using a genetic algorithm approach. SPIE Vol. 1006. Space Station Automation IV, pp. 226–234.

Bainbridge, B (1988). The explicit representation of control knowledge. In: Ringland, GA and Duce, DA (Eds), Approaches to Knowledge Representation: An Introduction. Research Studies Press, Letchworth, 260pp.

Baker, KD, Ball, LJ, Culverhouse, PF, Dennis, I, Evans, J, Jagodzinski, AP, Pearce, PD, Scothern, DGC and Venner, GM (1989). A process oriented approach to an intelligent design aid. Association for Computer Machinery, pp. 479–485.

Balachandran, M, Rosenman, MA and Gero, JS (1991). A knowledge based approach to the automatic verification of designs from CAD databases. Proc. AI in Design '91. Butterworth-Heinemann.

Bhatnagar, RK and Kanal, LN (1986). Handling uncertain information. A review of numeric and non-numeric methods. In: Kanal, LN and Lemmer, JF (Eds), Uncertainty in Artificial Intelligence. North Holland, pp. 3–26.

Beitz, W (1987). Designing for ease of assembly. In: Eder, WE, ICED-87. ASME, New York, pp. 767–773.

Bennett, JS and Englemore, RS (1979). SACON: a knowledge based consultant for structural analysis. Sixth International Joint Conference on AI, 47–49.

Black, WJ (1986). Intelligent Knowledge Based Systems: An Introduction. Van Nostrand Reinhold, London.

Bobrow, DG and Winograd, T (1977). An overview of KRL. Cognitive Science, 1, pp. 3–46.

Bobrow, DG and Winograd, T (1979). KRL: another perspective. Cognitive Science, 3, pp. 29–42.

Bond, A (1988). The cooperation of experts in engineering design. In: Gasser, L and Huhns, MN (Eds), Distributed Artificial Intelligence, Vol 2, Pitman, London, pp. 463–483.

Bonnissone, PP (1987). Plausible reasoning. In: Shapiro, SC (Ed), Encyclopedia of Artificial Intelligence: 2. Wiley, New York.

Boose, JH (1986). Rapid acquisition and combination of knowledge from multiple experts in the same domain. Future Computing Systems, 1, pp. 191–216.

Brachman, RJ (1979). On the epistomological status of semantic networks. In: Findler, NV (Ed), Associative Networks: Representation and Use of Knowledge by Computers. Academic Press, New York, pp. 3–50.

Brachman, RJ (1985). I Lied About the Trees Or, Defaults and Definitions in Knowledge Representation. Morgan Kaufman, Los Altos, California.

Brachman, RJ and Levesque, HJ (1985). Readings in Knowledge Representation. Morgan Kaufmann, Los Altos, California.

Bramer, M (1989). Expert systems: where are we and where are we going? In: Forsyth, R (Ed), Expert Systems: Principles and Case Studies, 2nd edn. Chapman and Hall, London, pp. 31–52.

Bramer, M (1990). Practical Experiences in Building Expert Systems. Wiley, New York.

Bratko, I (1986). PROLOG: Programming for Artificial Intelligence. Addison Wesley.

Breuker, JA and Wielinga, BJ (1987). Use of models in the interpretation of verbal data. In: Kidd, AL (Ed), Knowledge Acquisition for Expert Systems – A Practical Handbook. Plenum, New York, Chapter 2, pp. 17–44.

Broadbent, GH (1966). Creativity. In: Gregory, SA (Ed), The Design Method, Butterworths, London, 111–119.

Brouwer-Janse and Pitt (1986). Knowledge Acquisition: Methodological Issues and Problem Solving Profiles. Proc. of ECAI, Vol. 2, pp. 120–127.

Brown, DC, 1984. Expert systems for design problem solving using design refinement with plan selection and redesign, PhD thesis, Ohio State University.

Brown, DC and Chandrasekaran, B (1983). An approach to expert systems for mechanical design. IEEE Comp. Soc., NBS, Gaithersburg, Maryland, pp. 173–180.

Brown, DC and Chandrasekaran, B (1985). Expert systems for a class of mechanical design activity. In Gero, JS (Ed), Knowledge Engineering in CAD, Elsevier, Oxford, pp. 259–282.

Brown, DC and Chandrasekaran, B (1989). Design Problem Solving: Knowledge Structures and Control Strategies. Morgan Kaufmann, Los Altos, California.

Bucciarelli, LL (1988). An ethnographic perspective on engineering design, Design Studies, 9 (3), 159–168.

Buchanan, BG and Duda, RO (1983). Principles of rule based expert systems. In: Yovits, MC (Ed), Advances in Computers 22. Academic Press, New York.

Buchanan, BG, Sutherland, GL and Feigenbaum, EA (1969). Rediscovering some problems of artificial intelligence in the context of organic chemistry. In: Meltzer, B and Mitchie, D (Eds), Machine Intelligence: 5. Edinburgh University Press, pp. 253–280.

Buchanan, BG, Barstow, D, Bechtel, R, Bennett, J, Clancey, W, Kulikowski, C, Mitchell, T and Waterman, DA (1983). Constructing an expert system. In: Hayes-Roth, F, Waterman, DA and Lenat, B (Eds), Building Expert Systems. Addison Wesley, New York, pp. 127–168.

Buchanan, BG et al. (1983b). Constructing an expert system. In: Hayes-Roth, F, Waterman, DA and Lenat, DB (Eds), Building Expert Systems, Addison Wesley, New York.

Burgoyne, CJ and Jayasinghe, MTR (1991). Expert systems for structural design – not yet the whole solution, Presented at a meeting of the Institute of Structural Engineers, June, 1991, London, 17pp.

Burgoyne, CJ and Sham, SHR (1987). Application of expert systems to prestressed concrete bridge design. Civ. Eng. Systems, 4, pp. 14–19.

Camacho, G (1985). LOW-RISE: An Expert System for Structural Planning and Design of Industrial Buildings. Master's Thesis, Department of Civil Engineering, Carnegie Mellon University, Pittsburgh, Pennsylvania.

Carroll, DW (1985). Programming with Turbo Pascal. McGraw Hill/ MicroText. New York.

Castillo, E and Alvarez, E (1991). Uncertainty and Learning. Elsevier Applied Science.

Chabris, CF (1988). A Primer of Artificial Intelligence, Kogan Page, London.

Chan, WT and Paulson, BC (1987). Exploratory design using constraints. AI EDAM, 1, 59–71.

Chandrasekaran, B (1983). Towards a taxonomy of problem solving types. AI Magazine, 4 (1), 9–17.

Chandrasekaran, B and Mittal, S (1983). Deep versus compiled knowledge: approaches to diagnostic problem solving. Int. J. Man Machine Studies, 19, pp. 425–436.

Chandrasekaran, B and Tanner, MC (1986). Uncertainty Handling in Expert Systems: Uniform vs. Task Specific Formalisms. In: Kanal, LN and Lemmer, JF (Eds), Uncertainty in Artificial Intelligence. North Holland, Amsterdam, pp. 35–46.

Chang, TC, Ibbs, CW and Crandall, KC (1988). A fuzzy logic system for expert systems. Artificial Intelligence for Engineering Design, Analysis and Manufacturing, 1 (1).

Chatalic, P, Dubois, D and Prade, H (1987). An Approach to Approximate Reasoning Based on the Dempster Rule of Combination. Int. J. Expert Sys., 1 (1), pp. 67–87.

Cheeseman, P (1986). Probabilistic vs fuzzy reasoning. In: Kanal, LN and Lemmer, JF (Eds), Uncertainty in Artificial Intelligence. North Holland, Amsterdam, pp. 85–102.

Chehayeb, F and Connor, JJ (1985). GEPSE – A computer Environment for Engineering Problem Solving, MIT Research Report R86-11, Department of Civil Engineering, MIT.

Chehayeb, FS, Connor, JJ and Slater, JH (1985). An environment for building engineering KBS. In: Dym, CL (Ed), Applications of KBS to Engineering Design and Analysis. ASME, New York, AD-10, 9–28.

Choi, KC and Ibbs, CW (1990). Costs and benefits of computerisation in design and analysis. J. Comp. Civ. Engng, 4 (1), ASCE, 91–106.

Chung, P and Inder, R (1990). Clever computers: why should engineers use AI?, IEE Review, May, 189–193.

Chung, PWH and Kumar, B (1987). Knowledge elicitation methods: a case study in structural design. In: The Application of Artificial Intelligence Techniques to Civil and Structural Engineering (Conf.), Civil-Comp Press, Edinburgh, pp. 21–26.

Clancey, WJ (1981). Methodology for Building an Intelligent Tutoring System. Computer Science Dept Report No. STAN-CS-81-894, Stanford University, Palo Alto, California.

Clancey, W (1983). The epistomology of rule based systems: a framework for explanation. In: Brachman, R and Levesque, H (Eds), Readings in Knowledge Representation. Morgan Kaufman, Los Altos, California.

Clausing, DP (1985). Product Development Process, IEEE, CH2175-8, 896–900.

Clausing, DP and Ragsdell, KM (1985). The efficient design of medium and light machinery using state-of-the-art technology. In: Yoshikawa, H (Ed), Design and Synthesis, Elsevier, Amsterdam, 203–208.

Clayton, BD (1985). ART Programming Primer and Programming Tutorial, Vols. 1, 2 and 3. Inference Corp.

Clitheroe, Bellcore and Fisher (1989). Knowledge based assistance of genetic search in large design spaces. Association for Computing Machinery, 729–733.

Clocksin, WF and Mellish, CS (1984). Programming in PROLOG. Springer-Verlag, Berlin.

Cohen, PR (1985). Heuristic Reasoning about Uncertainty: An Artificial Intelligence Approach. Pitman, London.

Cohn, LF, Harris, RA and Bowlby, W (1988). Knowledge acquisition for domain experts, J. Comp. Civ. Engng, 2 (2), ASCE, 107–120.

Colgan, L and Spence, R (1991). Cognitive modelling of electronic design. Proc. AI in Design '91. Butterworth-Heinemann, London.

Collins, HM (1990). Artificial Experts: Social Knowledge and Intelligent Machines. MIT Press, Cambridge, Massachusetts, 266pp

Colmerauer, A (1987). Opening the PROLOG III universe. BYTE Magazine, August.

Colmerauer, A (1990). An introduction to PROLOG III Communication of ACM, 33, July.

Conway, T and Wilson, M (1988). Psychological studies of knowledge representation. In: Ringland, GA and Duce, DA (Eds.), Approaches to Knowledge Representation: An Introduction. Research Studies Press, Letchworth, pp. 117–160.

Coyne, RD, Rosenman, MA, Radford, AD et al., (1990). Knowledge based design systems. Addison-Wesley, Reading, Massachusetts, 567 pp

Davis, R (1980). Meta rules: reasoning about control. AI, 15, pp. 179–222.

Davis, R (1982). TEIRESIAS: applications of meta level knowledge. Part 2 of Davis, R and Lenat, DB, Knowledge Based Systems in Artificial Intelligence. McGraw-Hill, New York.

Davis, R and King, J (1977). An Overview of Production Systems. In: Elcock, E and Mitchie, D (Eds): Machine Intelligence: 8. Ellis Horwood, Chichester.

De Millo, RA, Lipton, RJ and Perlis, AJ (1979). Social processes and proofs of theorems and programs. Communications of ACM, 22 (5), pp. 271–280.

Dempster, AP (1968). A generalisation of Bayesian Inference. Journal of the Royal Statistical Society, 30.

Dincbas, M (1988). Constraints, logic programming and deductive databases. In: Fuchi, K and Nivat, M (Eds), Programming of Future Generation Computers. Elsevier Science BV, North-Holland.

Dincbas, M, Van Hentenryck, P, Simonis, H, Aggoun, A and Graf, T (1988). Applications of CHIP to industrial and engineering problems. First International Conference on Industrial and Engineering Applications of AI and Expert Systems, Tullohoma, Tennessee, June.

Dincbas, M, Van Hentenryck, P, Simonis, H, Aggoun, A and Berhier, F (1988). The constraint logic programming language CHIP Proceedings of the International Conference on Fifth Generation Computer Systems, Tokyo, pp. 693–702.

Dincbas, M, Simonis, H and Van Hentenryck, P (1990). Solving large combinatorial problems in logic programming. Journal of Logic Programming, 8, pp. 75–93.

Dixon, JR (1966). Design Engineering: Inventiveness, Analysis and Decision Making, McGraw-Hill, New York, 354pp.

Dixon, JR and Simmons, MK (1983). Computers that design: expert systems for mechanical engineers. Computers in Mechanical Engineering, pp. 10–18.

Doyle, J (1983). Methodological simplicity in expert system construction: the case of judgements and reasoned assumptions. AI Magazine, 4 (2), pp. 39–43.

Drach, A, Langenegger, M and Heitz, S (1992). Flexible Environments for Integrated Building Design, CAD92, 19pp.

Duce, DA and Ringland, GA (1988). Background and introduction. In: Ringland, GA and Duce, DA (Eds.), Approaches to Knowledge Representation: An Introduction. Research Studies Press, Letchworth, pp. 1–12.

Duda, J, Gaschnig, J and Hart, P (1979). Model design in the prospector consultant system for mineral exploration. In: Mitchie, D (Ed), Expert Systems in the Microelectronic Age. Edinburgh University Press, Edinburgh, pp. 153–167.

Dyer, MG, Flowers, M and Hodges, J (1986). EDISON: an engineering design invention system operating naively. In: Sriram, D and Adey, R (Eds), Proc. App. AI in Engng Probs. Springer-Verlag, New York.

Dym, CL and Levitt, RE (1991). Knowledge-Based Systems in Engineering, McGraw-Hill, New York, 404pp.

Eastman, CM (1978). The representation of design problems and maintenance of their structure. In: Latcombe, JC (Ed), AI and Pattern Recognition in CAD, IFIP, North Holland, Amsterdam, 335–365.

Eastman, CM (1981). Recent developments in representation in the science of design. Proceedings of the Eighteenth Design Automation Conference, ACM, IEEE, pp. 12–21.

Eder, WE (1966), Definitions and methodologies. In: Gregory, SA (Ed), The Design Method, Butterworths, London, 19–31.

Eder, WE (1987). Survey of design education. In: Eder, WE (Ed), ICED-87. ASME, New York, pp. 881–887.

Ehrlenspiel, K (1987). Reduction of Product Costs in West Germany. In: Eder, WE (Ed), ICED-87. ASME, New York, pp. 796–806.

Ehrlenspiel, K and John, T (1987). Inventing by design methodology. In: Eder, WE (Ed), ICED-87. ASME, New York, pp. 29–37.

Eisenberg, J and Hill, J (1984). Using natural language systems on personal computers. BYTE, January, pp. 226–238.

Ellman, J (1987). Practical experiences of knowledge acquisition for expert systems. In: Pavelin, CJ and Wilson, MD (Eds), Proceedings of an SERC Workshop on Knowledge Acquisition for Engineering Applications, 1987, pp. 39–43.

Ericsson, K and Simon, HA (1980). Verbal reports as data. Psychological Review, 8 (3), pp. 215–251.

Esterline, A, Rosen, D, Otto, K, Nelson, L, Hessburg, T, Riley, DR and Erdman, AG (1988). A methodology for capturing mechanical design expertise. Computers in Engineering, Proeedings of the ASME Conference, San Francisco, USAASME, New York, 47–55.

Etter, DM and Dayton, DC (1983). Performance characteristics of a genetic algorithm in adaptive IIR filter design. Signal processing ii: Theories and applications – Proceedings of the EUSIPCO-83 Second European Signal Processing Conference, pp. 53–56.

Evans, JSt.BT (1989). Bias in Human Reasoning: Causes and Consequences, Lawrence Erlbaum Assocs, Hove, UK, 200pp.

Fair, GM, Geyer, JC and Okun, DA (1971). Elements of Water Supply and Wastewater Disposal, Wiley, New York, 752pp.

Feigenbaum, EA (1982). Knowledge Engineering for the 1980's. Computer Science Dept, Stanford University, Stanford, California.

Feigenbaum, EA, Buchanan, B and Lederburg, J (1971). On generality and problem solving: a case study involving the DENDRAL program. Machine Intelligence, 6.

Feigenbaum, EA and McCorduck, P (1983). The Fifth Generation. Addison Wesley, New York.

Fenves, SJ (1985). A framework for a KB finite element analysis assistant. In Dym, CL (Ed), Applications of KBS to Engineering Design and Analysis. ASME, New York, AD-10, 1–8.

Feyock, S and Rogers, JL (1987). Adding intelligence to a database management system. In: Sriram, D and Adey, RA (Eds), AI in Engineering: Tools and Techniques, 177–195.

Finger, S and Dixon, JR (1989a). A review of research in mechanical engineering design. Part 1: descriptive, prescriptive and computer-based models of design processes. Research in Engineering Design, 1, Springer-Verlag, New York, 51–67.

Finger, S and Dixon, JR (1989b). A review of research in mechanical engineering design. Part II: representations, analysis and design for the life cycle, Research in Engineering Design, 1, 121–127.

Fisher, DC and Nguyen, TD (1989). Design and analysis aid for evaluating aircraft structures. ACM, 469–472.

Fitzpatrick, JM, Gerefstette, JJ and Van Gucht, D (1984). Image registration by genetic search. Conference Proceedings of the IEEE, Soiteastcon 84, pp. 460–464.

Flemming, U (1987). The role of shape grammars in the analysis and creation of designs. In: Kalay, YE (Ed), Computability of Design. Wiley, New York, 245–272.

Ford, N (1989). PROLOG Programming. Wiley.

Forgy, C (1982). RETE: A fast algorithm for the many pattern/many object pattern match problem, Artificial Intelligence, 19, pp. 17–37.

Forsyth, R (Ed) (1984). Expert Systems: Principles and Case Studies, 1st edn. Chapman and Hall, London.

Forsyth, R (Ed) (1989). Expert Systems: Principles and Case Studies, 2nd edn. Chapman and Hall, London.

Forsyth, R (1989). The expert systems phenomenon. In: Forsyth, R (Ed). Expert Systems: Principles and Case Studies, 2nd edn. Chapman and Hall, London, pp. 3–20.

Fox, J (1986). Knowledge, decision making and uncertainty. In: Gale, W (Ed), Artificial Intelligence and Statistics. Addison Wesley, New York, pp. 57–76.

Fox, J (1987). Dealing with uncertainty. In: O'Shea, T, Self, J and Thomas, G (Eds.), Intelligent Knowledge Based Systems. Harper and Row, pp. 52–67.

Fox, J, Myers, CD, Greaves, MF and Pegram, S (1987). In: Kidd, AL (Ed), Knowledge Acquisition for Expert Systems: A Practical Handbook. Plenum Press, New York. pp. 73–90.

Fox, MS et al. (1983). Techniques for sensor based diagnosis. Proceedings of the Eighth International Joint Conference on AI, pp. 158–163.

Franck, BM (1989). Qualitative engineering at various levels of conception for design and evaluation of structures. Association of Computer Machinery, 441–448.

Fraser, DJ (1981). Conceptual Design and Preliminary Design of Structures. Pitman, London.

Freeman, P and Newell, A (1971). A Model for functional reasoning in design. In: IJCAI 2, pp. 621–640.

French, MJ (1985). Conceptual Design for Engineers, 2nd edn. The Design Council, London, 236pp.

Frost, RA (1986). Introduction to Knowledge Based Systems. Collins, London.

Fukuda, S (1988). Codes and rules and their roles as constraints in expert systems for structural design. In: Adeli, H (Ed), Expert Systems in Construction and Structural Engineering. Chapman and Hall, London, pp. 309–323.

Furse, G (1989). Debugging Knowledge Bases. In: Forsyth, R (Ed), Expert Systems: Principles and Case Studies, 2nd edn. Chapman and Hall, London, pp. 184–196.

Gammack, JG (1987). Different techniques and different aspects on declarative knowledge. In: Kidd et al. (Eds), Knowledge Acquisition for Expert Systems – A Practical Handbook. Plenum: New York. Chapter 7, pp. 137–162

Garavaglia, S (1987) PROLOG: Programming Techniques and Applications. Harper and Rowe, London.

Gaschnig, J (1979). Preliminary performance analysis of the prospector consultant system for mineral exploration. Proceedings of the Sixth International Joint Conference on AI, Morgan Kaufmann, Los Altos, California, pp. 308–310.

Gaschnig, J (1980). Development of uranium exploration models for the PROSPECTOR consultant system. SRI Project 7856, AI Centre, SRI International, Menlo Park, California, USA.

Gaschnig, J, Klhar, P, Pople, H, Shortliffe, E and Terry, A (1983). Evaluation of expert systems: evaluations and case studies. In: Hayes-Roth, F and Waterman, D (Eds), Building Expert Systems. Addison Wesley, New York, pp. 241–282.

Gauchel, J, Drach, A and Hovestadt, L (1990). Ein intelligentes Planungswekzeug fur das Verlegen haustechnischer leitungen in Hochinstallierten gebauden. VDI Berichte, 861 (5), pp. 39–54.

Gero, JS (Ed) (1988). AI in Engineering: Design, Elsevier, Amsterdam, 465pp.

Gero, JS and Maher, ML (1988). Designing with Prototypes. ICED-88, Budapest, Hungary, pp. 74–81.

Gero, JS, Maher, ML and Zhang, W (1988). Chunking structural design knowledge as prototypes. In: Third International Conference on AI in Engineering. 3–21.

Gero, JS and Stanton, R (Eds) (1988). Artificial Intelligence Developments and Applications. North Holland.

Gill, H (1987). Design for manufacture – a case study. In: Eder, WE (Ed), ICED-87. ASME, New York, pp. 807–814.

Goel, A and Chandrasekaran, B (1989). Functional representation of designs and redesign problem solving. Proc. UCAI-89, pp. 1388–1394.

Goldberg, DE (1989). Genetic Algorithms in Search, Optimisation and Machine Learning, Addison Wesley, Reading, Massachusetts.

Goodall, A (1989). An introduction to expert systems. In: Forsyth, R (Ed), Expert Systems: Principles and Case Studies, 2nd Edn. Chapman and Hall, London, pp. 22–29.

Graham, I (1989). Inside the inference engine. In: Forsyth, R (Ed), Expert Systems: Principles and Case Studies, 2nd edn. Chapman and Hall, London, pp. 57–82.

Gray, C and Little, J (1985). A systematic approach to the selection of an appropriate crane for a construction site. Construction Management and Economics, Vol. 3, pp. 121–144.

Gray, PMD (1985). Expert systems and object oriented databases: evolving a new software architecture. Research and development in expert systems, V Proceedings of Expert Systems '88: The Eighth Annual Technical Conference of the British Computer Society, 12–15 December

Greenwell, M (1987). Knowledge Engineering For Expert Systems. Ellis Horwood, Chichester.

Gregory, SA (Ed) (1966). The Design Method, Butterworths, London, 350pp.

Gregory, SA (1987). Expert system versus creativity in design. In: Eder WE (Ed), ICED-87. ASME, New York, pp. 615–621.

Grimson, WEL and Patil, RS (Eds) (1986). AI in the 1980's and Beyond: An MIT Survey. MIT Press, Cambridge, Massachusetts.

Grinberg, MR (1980). A knowledge based system for digital electronics. First Annual National Conference on AI, Stamford, California, pp. 283–286.

Grover, MD (1983). A pragmatic knowledge acquisition methodology. Proceedings of the Eighth International Conference on AI, pp. 436–438.

Gupta, MM and Yamakawa, T (Eds) (1988). Fuzzy Logic in Knowledge Based Systems and Control. North Holland, Amsterdam.

Hammond, KJ (1986). CHEF: a model of a case based planner. AAAI, 86, Philadelphia, Pennsylvania, 267–271.

Hansen, JV and Messier, WF (1985). A Preliminary Investigation of EDP-Xpert. Working Paper 85-6. Accounting Research Center, University of Florida.

Harris, LR (1977). User oriented data base query with the ROBOT natural language query system. Proceedings of the Third International Conference on Very Large Databases. Tokyo.

Hart, A (1986). Knowledge Acquisition for Expert Systems. Kogan Page, London.

Hartley, RT (1982). A conceptual basis for expert systems methodology. Proceedings of the Second British Computer Society Conference on Expert Systems, Brunel University, Surrey, pp. 47–51.

Hartnell, T (1984). Exploring Artificial Intelligence. Interface Publications, London.

Hayes, PJ (1977). In defense of logic. Proceedings of the Fifth IJCAI Morgan Kaufmann, Los Altos, California, pp. 559–565.

Hayes, PJ (1979). The logic of frames. In: Metzing, D (Ed), Frame Conception and Text Understanding. Walter de Gruyter, Berlin.

Hayes-Roth, B (1984). A Blackboard Model of Control. Report No. HPP-83-38. Computer Science Dept., Stanford University

Hayes-Roth, F, Waterman, D and Lenat, D (Eds) (1983). Building Expert Systems. Addison-Wesley, Reading, Massachusetts.

Hayward, ACG (1989a). Internal Report. Cass Hayward and Partners, Chepstow, Gwent.

Hayward, ACG (1989b). Composite Steel Highway Bridges. British Steel.

Hayward, A (1990). Personal Communication. Senior Partner, Cass Hayward and Partners, Chepstow, Gwent.

Hendrix, G (1979). Encoding knowledge in partitioned networks. In: Findler, NV (Ed), Associative Networks: Representation and Use of Knowledge by Computer. Academic Press, New York.

Hetzel, WC (1973). Principles of computer testing. In: Hetzel, WC (Ed), Program Test Methods. Prentice Hall, New Jersey, pp. 17–28.

Hickam, DH et al. (1985). The treatment advice of a computer based cancer chemotherapy protocol advisor. Annals of Internal Medicine, 103 (6), Part 1, pp. 928–936.

Holland, JH (1986). Escaping brittleness: the possibilities of general purpose learning algorithms applied to parallel rule based systems. In: Michalski, RS, Carbonell, JG and Mitchell, TM (Eds), Machine Learning: An Artificial Intelligence Approach. Vol. II Morgan Kaufmann Publishers, Los Altos, pp. 593–623.

Hooper, JN, Barclay, AR and Miles, JC (1992). Increasing the Reliability and Convergence of a Genetic Algorithm in a Varying Scale Multi-objective Engineering Problem. Institution of Electrical Engineers (London), Digest 1992/011, 2/1-2/5.

Horvitz, E and Heckerman, D (1986). The inconsistent use of measures of uncertainty in artificial intelligence research. In: Kanal, LN and Lemmer, JF (Eds), Uncertainty in Artificial Intelligence. North Holland, Amsterdam, pp. 137–152.

Huang, GQ (1990). Cooperating Knowledge Based Systems for Manufacturing Design, PhD thesis, School of Engineering, University of Wales, Cardiff, 300pp.

Huang, GQ and Brandon, JA (1992). AGENTS: object-oriented prolog system for cooperating knowledge-based systems. Knowledge Based Systems, 5 (2), 125–136.

Huber, GP (1974). Methods for quantifying subjective probabilities and multi-attribute utilities. Decision Sciences, 5, pp. 431–458.

Hubka, V (1982). Principles of Engineering Design. Translated by Eder, WE, Butterworths, London, 118pp.

Hubka, V (1987). A curriculum model applying the theory of technical systems. In: Eder, WE (Ed), ICED-87. ASME, New York, pp. 965–976.

Hubka, V and Schregenberger, JW (1987). Paths towards design science. In: Eder, WE (Ed), ICED-87. ASME, New York, pp. 3–14.

Hudson, DL et al. (1984). Prospective analysis of emerge, an expert system for chest pain analysis. In IEEE Computers in Cardiology, IEEE Service Centre, New Jersey, pp. 19–24.

Hughes, S (1986). Question classification in rule base systems. In: Bramer, M (Ed), Research and Development in Expert Systems III: Cambridge University Press, Cambridge.

Hundal, MS (1987). Conceptual Design of Power Transmissions. In: Eder, WE (Ed), ICED-87. ASME, New York, pp. 246–253.

Hunter, AB (1991). Developments in Artificial Intelligence Reasoning. In: Winstanley, G (Ed), Artificial Intelligence in Engineering. Wiley.

Hutchinson, PJ, Rosenman, MA and Gero, JS (1987). RETWALL: an expert system for the selection and preliminary design of earth retaining structures, Knowledge Based Systems, 1 (1), 11–23.

Inder, R, Chung, PWH and Fraser, J (1987). Towards a methodology for incremental system development. In: Pavelin, CJ and Wilson, MD (Eds), Proceedings of a SERC Workshop on Knowledge Acquisition for Engineering Applications. pp. 44–55

Ishii, K and Hornberger, L (1991). Keys to the successful development of an AI based tool for life-cycle design. Proc. AI in Design '91. Butterworth-Heinemann, London.

Jaffer, J and Lassez, JL (1987). Constraint logic programming. In: Proceedings of the Fourteenth ACM Symposium on Principles of Programming Languages, Munich.

Johnson, RP and Buckby, RJ (1986). Composite Structures of Steel and Concrete. Collins, London.

Johnson-Laird, PN and Watson, P (Eds) (1975). Thinking: Readings in Cognitive Science. Cambridge University Press Cambridge.

Jones, JC (1980). Design Methods: Seeds of Human Futures. Wiley, New York, 407pp.

Jones, P (1984). REVEAL: an expert systems support environment. In: Forsyth, R (Ed), Expert Systems, Principles and Case Studies, 1st edn. Chapman and Hall, London, pp. 133–150.

Jones, P (1989). Uncertainty management in expert systems. In: Forsyth, R (Ed), Expert Systems: Principles and Case Studies. Chapman and Hall, London, pp. 106–120.

Kalay, YE (1987). Computability of Design. Wiley, New York, 363pp.

Keen, M (1990). Expert systems in clarifying employment law. In: Bramer, M (Ed), Practical Experience in Building Expert Systems. pp. 75–102.

Kelly, A and Pohl, I (1986). A Book on C: An Introduction to Programming in C, 1st edn. Benjamin/Cummings, Redwood City, California.

Kelly, A and Pohl, I (1990). A Book on C: Programming in C, 2nd edn. Benjamin/Cummings, Redwood City, California.

Kelly, GA (1955). The Psychology of Personal Constructs. Norton, New York.

Keravnou, ET and Johnson, L (1986). Competent Expert Systems. Kogan Page, London.

Kerley, JJ (1987). Retroduction: a new structured approach to mechanical design. In: Eder, WE (Ed), ICED-87. ASME, New York, pp. 123–138.

Kernighan, BW and Ritchie, DM (1988). The C Programming Language, 2nd edn. Prentice Hall, New Jersey.

Kidd, AL (1985). What do users ask – some thoughts on diagnostic advice. In: Merry, M (Ed), Expert Systems '85. Cambridge University Press, Cambridge, pp. 9–19.

Kidd, AL (1987). Knowledge acquisition: an introductory framework. In Kidd, AL (Ed), Knowledge Acquisition for Expert Systems: A Practical Handbook. Plenum Press, New York, pp. 1–16.

Kidd, AL and Cooper, MB (1983). Man machine interface for an expert system. Third British Computer Society Conference on Expert Systems, Cambridge University, Cambridge, UK.

Kitzmiller, CT and Kowalik, JS (1987). Coupled symbolic and numeric computing in knowledge based systems. AI Magazine, 85–90.

Klahr, P and Waterman, DA (1986). Expert System Techniques, Tools and Applications. Addison Wesley, Reading, Massachusetts.

Kolodner, JL (1991). Improving human decision making through case based decision aiding. AI Magazine, 12 (2), Summer.

Kolodner, JL, Simpson, RL and Sycara-Cyranski, K (1985). A process model of case based reasoning in problem solving. Ninth International Conference on AI, 284–290.

Kostem, CN (1986). Design of an expert system for the rating of highway bridges. In: Expert Systems in Civil Engineering. ASCE, New York.

Krieger, MH (1987). Learning to break the rules: what makes engineers (and other professionals) successful. In: Eder, WE (Ed), ICED-87. ASME, New York, pp. 927–935.

Kulikowski, CA and Weiss, SH (1982). Representation of expert knowledge for consultation: the casnet and expert projects. In: Szolovits, P (Ed), Artificial Intelligence in Medicine. West View Press, pp. 21–56.

Kumar, B and Topping, BHV (1989). An integrated rule-based system for industrial building design. In: Adeli, H (Ed), Microcomputer Knowledge Based Expert Systems in Civil Engineering. ASCE, New York.

Lange, R (1987). Exception hierarchies as a knowledge representation for expert systems. In: Sriram, D and Adey, RA (Eds), AI in Engineering: Tools and Techniques, 1–10.

Lansdown, J (1989). The designers' information environment: some tools for design knowledge manipulation. Civ. Eng. Sys, 5–10.

Latcombe, JC (1979). Failure processing in a system for designing complex assemblies. Sixth International Joint Conference on AI, Tokyo, 508–515.

Lenat, DB (1982). The nature of heuristics. Artificial Intelligence, 19 (2), 189–240.

Levitt, RE and Kunz, JC (1987). Using artificial intelligence techniques to support project management. Journal of Artificial Intelligence Techniques in Engineering Design, Analysis and Manufacturing. 1 (1).

Lewis, D (1973). Counterfactuals. Harvard University Press, Cambridge, Massachusetts.

Lindley, DV (1987). The probability approach to the treatment of uncertainty. In: Artificial Intelligence and Expert Systems. Statistical Science. 2 (1), pp. 17–24.

Lindsay, PH and Norman, DA (1972). Human Information Processing. Academic Press, New York, 500pp.

Lindsay, PH and Norman, DA (1977). Human Information Processing, 2nd edn. Academic Press, New York.

Lindsay, R, Buchanan, BG, Feigenbaum, EA and Lederberg, J (1981). Applications of AI for Organic Chemistry: The DENDRAL Project. McGraw-Hill, New York.

Mackenzie, CA and Gero, JS (1987). Learning design rules from decisions and performances. Int. J. Artific. Intel. Engng, 2 (1), 2–11.

MacRandal, D (1988). Semantic networks. In: Ringland, GA and Duce, DA (Eds), Approaches to Knowledge Representation: An Introduction. Research Studies Press, Letchworth, pp. 45–80.

Maher, ML (1984). A knowledge-based expert system for the preliminary design of high rise buildings. PhD thesis, Department of Civil Engineering, Carnegie-Mellon University, Pittsburgh, Pennsylvania.

Maher, ML (Ed) (1987). Expert Systems for Civil Engineers. ASCE, New York.

Maher, ML and Allen, RA (1987). Expert System Components. In: Maher, ML (Ed), Expert Systems for Civil Engineers: Technology and Application. ASCE, pp. 3–13.

Maher, ML and Zhang, DM (1991). CADSYN: using case and decomposition knowledge for design synthesis. Proc. AI in Design. Butterworth-Heinemann, London.

Malpas, J (1988). PROLOG: A Relational Language and its Applications. Prentice Hall, New Jersey.

Marples, DL (1961). Decisions of engineering design. IRE Transactions on Engineering Management, EM (8), 55–71.

Martin, J and Oxman, S (1988). Building Expert Systems: A Tutorial. Prentice Hall Press, Edinburgh, 199–203.

McCarthy, J et al. (1965). LISP 1-5, Programmers Manual, 2nd edn. MIT Press, Cambridge, Massachusetts.

McCarthy, TJ and Nouas, Z (1991). A knowledge representation scheme for the design of hybrid structural steel work connections. In: Topping, BHV (Ed), AI and Structural Engineering. CIVIL-COMP.

McCrory, RJ (1966). The design method in practice. In: Gregroy, SA (Ed), The Design Method, Butterworths, London, 11–18.

McDermott, D (1987). A critique of pure reason. Journal of Computational Intelligence.

McDermott, J (1980). R1's formative years. AI Magazine, 2 (2), pp. 21–29.

McDermott, J (1982). R1: A rule-based configurer of computer systems. Artificial Intelligence, 19 (1).

McDonald, JE, Dearholt, DW, Paap, KR and Schvaneveldt, RW (1986). A formal interface design methodology based on user knowledge. Proceedings of Human Factors in Computing Systems. Association for Computing Machinery, pp. 285–290.

McGraw, KL and Seale, MR (1988). Knowledge elicitation with multiple experts. AI Review, 2 (1).

McIntosh, PG (1987). Models of spatial information in computer-aided architectural design: a comparative study. In: Kalay, YE (Ed), Computability of Design. Wiley, New York, pp. 117–131.

Miles, JC and Moore, CJ (1989). An expert system for the conceptual design of bridges. In: Topping, BHV (Ed), AI Techniques and Applications for Civil and Structural Engineers. CIVIL-COMP Press, Edinburgh, pp. 171–176.

Miles, JC and Moore, CJ (1991). Conceptual design: pushing back the boundaries using knowledge based systems. In: Topping, BHV (Ed), AI and Structural Engineering. CIVIL-COMP Press, Edinburgh, 73–78.

Miles, JC, Moore, CJ and Hooper, JN (1990). A structured multi-expert knowledge elicitation methodology for the development of practical knowledge based systems. Institute of Electrical Engineers (London), Computing and Control Division, Digest 1990/077, 3pp.

Miller, RA, Pople, HE and Myers, JD (1982). Internist-1: an experimental computer based diagnostic consultant for general internal medicine. The New England Journal of Medicine, 307 (8), pp. 468–476.

Milne, R (1990). Intelligent data interpretation. In: Bramer, M (Ed), Practical Experience in Building Expert Systems. Wiley, Chichester, pp. 103–124.

Minsky, M (1975a). A Framework for Representing Knowledge. In Winston, PH (Ed). The Psychology of Computer Vision. McGraw-Hill, New York.

Minsky, M (1975b). Frame system theory. In: Johnson Laird, PN and Wason, P (Eds), Thinking: Readings in Cognitive Science. Cambridge University Press, Cambridge.

Mitchell, TM, Steinberg, LI, Kedar-Cabelli, S, Kelly, VE, Schulman, J and Weinrich, T (1983). An intelligent aid for circuit design. Proceedings of the AAAI Conference, 274–278.

Mitchie, D (1980). Expert systems. Computer Journal, 23, 369–377

Mittal, S and Dym, CL (1985). Knowledge acquisition from multiple experts. AI Magazine, Summer.

Montgomery, A (1992). Combining SE discipline with AI creativity. IEE (London) Digest 1992/087, 1/1–1/5.

Moore, CJ (1991). An Expert System for the Conceptual Design of Bridges. Unpublished PhD Thesis, School of Engineering, Cardiff, Wales, February, 325 pp.

Moore, CJ and Miles, JC (1991a). Knowledge elicitation using more than one expert to cover the same domain. AI Review, 5, 255–271.

Moore, CJ and Miles, JC (1991b). The development and verification of a user oriented KBS for the conceptual design of bridges. Civil Engineering Systems, 8, 81–86.

Moore, CJ and Miles, JC (1991c). The importance of detailed KBS evaluation for systems intended for implementation in the engineering industry. Computing Systems in Engineering, 2 (4), 365–378.

Moore, CJ and Miles, JC (1992). In depth heuristic analysis in engineering design KBS. Institution of Electrical Engineers (London), Digest 1992/011, 6/1–6/4.

Moore, R (1985). The role of logic in knowledge representation and commonsense reasoning. In: Brachman, RJ and Levesque, HJ (Eds), Readings in Knowledge Representation. Morgan Kaufmann, Los Altos, California.

Morrison, D (1968). Engineering Design, McGraw-Hill, London, 202pp.

Morse, DV and Hendrikson, C (1991). Model for Communication in Automated Interactive Engineering Design. Journal of Computing in Civil Engineering. 5 (1), ASCE, 4–25.

Mostow, J (1985). Towards better models of the design process. AI Magazine, 6 (1), 44–57.

Mostow, J and Barley, M (1987). Automated Reuse of Design Plans. ICED-87, Boston, Massachusetts, pp. 632–647.

Mullarkey, PW (1987). Languages and tools for building expert systems. In: Maher, ML (Ed), Expert Systems for Civil Engineers. ASCE, New York, pp. 15–34.

Nash, GFJ (1984). Steel Bridge Design Guide – Composite Universal Beam Simply Supported Spans. Constrado, Croydon.

Navichandra, D and Sriram, D (1987). Analogy-based engineering problem solving: an overview. In: Sriram, D and Adey, RA (Eds), AI in Engineering: Tools and Techniques, 273–285.

Naylor, C (1983). Build Your Own Expert System, Sigma Technical, Chichester.

Nelson, J (1987). Advanced Graphics in C: Programming and Techniques. McGraw-Hill, New York.

Newell, A and Simon, HA (1972). Human Problem Solving. Prentice Hall, Englewood Cliffs, New Jersey.

Newsome, SL, Spillers, WR and Finger, S (1988). Design Theory '88. Proceedings of the 1988 NSF Grantee Workshop on Design Theory and Methodology. Springer-Verlag, New York, 365pp.

Nilsson, NJ (1980). Principles of AI, Toga Publishing Co.

Nilsson, NJ (1986). Probabilistic logic. Artificial Intelligence, 28, pp. 71–87.

Nutter, JT (1987). Uncertainty and probability. Proceedings of the Tenth International Conference on Artificial Intelligence, Milan. Morgan Kaufmann Publishers Inc., Los Altos, California.

O'Connor, C (1971). Design of bridge superstructures. Wiley-Interscience, New York.

Oherson, DN and Smith, EE (1981). On the adequacy of prototype theory as a theory of concepts. Cognition, 9, pp. 35–58.

O'Keefe, M., Balci, O and Smith, EP (1987). Validating expert system performance. IEEE, Winter, pp. 81–89.

Oxman, R and Gero, J (1987). Using expert systems for design diagnosis and design synthesis. Expert Systems, 4 (1), 4–15.

Paek, YJ and Adeli, H (1988). Representation of structural design knowledge in a symbolic language. JComp. Civ. Engng, 2 (4), 346–364.

Pahl, G and Beitz, W (1988). Engineering Design: A Systematic Approach, 2nd edn, translated into English by Pomerans, A and Wallace, K. Design Council, London, 397pp.

Parsaye, K and Chignell, M (1988). Expert Systems for Experts. Wiley, New York.

Partridge, D (1981). "Computational theorizing" as the tool for resolving wicked problems, IEEE Trans. on Systems, Man and Cybernetics, SMC-11 (4), 318–352.

Partridge, D (1985). The social implications of AIIn: Yazdani, M (Ed), AI: Implications and Applications. Chapman Hall, London.

Partridge, D (1988). Artificial Intelligence: Applications in the Future of Software Engineering. Ellis Horwood, Chichester.

Pavelin, C (1988). Logic in knowledge representation. In: Ringland, GA and Duce, DA (Eds), Approaches to Knowledge Representation: An Introduction (Eng. Library). Research Studies Press, Letchworth, pp. 13–44

Pecora, D, Zumsteg, JR and Crossman, FW (1985). An application of expert systems to composite structural design and analysis. In Dym, CL (Ed), Applications of KBS to Engineering Design and Analysis. ASME, New York, AD-10, 135–147.

Pham, DT (1988). Expert Systems in Engineering. IFS Publications, Springer-Verlag, London, 480pp.

Philbey, BT, Miles, C and Miles, JC (1991). The development of an interface for an expert system used for conceptual bridge design. In: Topping, BHV (Ed), AI and Structural Engineering, CIVIL-COMP Press, Edinburgh, 87–96.

Phillips, RE, Wilson, WW and Parmenter, K (1989). Applying qualitative knowledge to aircraft engine system design. Association for Computing Machinery, New York, pp. 473–478.

Pitts, G (1973). Techniques in Engineering Design. Butterworths, London, 173pp.

Pollock, J (1987). Defeasible reasoning. Synthese, 55, 231–252.

Porter, WL (1988). Notes on the inner logic of designing: two thought-experiments. Design Studies, 9 (3), 169–180.

Pu, P and Reschberger, M (1991). Assembly sequence using case based reasoning techniques. Proc. AI in Design '91. Butterworth-Heinemann, London.

Quillian, MR (1968). Semantic memory. In: Minsky, M (Ed), Semantic Information Processing. MIT Press, Cambridge, Massachusetts, pp. 216–270.

Rada, R, Rhine, Y and Smallwood, J (1984). Rule Refinement. Proceedings of the Eighth Annual Symposium on Computer Applications in Medical Care, pp. 62–65.

Radford, AD, Hung, P and Gero, JS (1984). New rules of thumb from computer aided structural design. Proceedings of the Sixth International Conference on Computers in Engineering Design. Guildford, UK, 558–566.

Ramstrom, D and Rhenman, E (1965). A method of describing the development of an engineering project. IEEE Trans. Eng. Management, EM-12 (3), 79–86.

Rasdorf, WJ and Parks, LM (1987). Natural language prototypes for analyzing design standards. In: Sriram, D and Adey, RA (Eds), AI in Engineering: Tools and Techniques, 145–160.

Rasdorf, WJ and Wang, TE (1988). Generic design standards processing in an expert system environment. J. Comp. Civ. Engng, 2 (1), 68–87.

Reboh, R (1979). The knowledge acquisition system. In: Duda, RO, Hart, PE, Konolige, K and Reboh, R (Eds), A Computer Based Consultant for Mineral Exploitation. Final Report, SRI Project 6415. AI Center, SRI Int., Menlo Park, California.

Rehak, DR (1985). An integrated knowledge based systems architecture for CASESIGART Newsletter, 91, 53–55.

Regh, J, Elfes, A, Talukdar, S, Eisenberger, M and Edahl, R (1988). CASE: computer aided simultaneous engineering. In: Gero, JS (Ed), AI in Engineering: Design, Elsevier, Amsterdam, pp. 339–359.

Rich, E (1983). Artificial Intelligence. McGraw-Hill, New York.

Richer, MH (1985). Evaluating the Existing Tools for Developing Knowledge Based Systems. Stanford Knowledge Systems Laboratory, 85, 19.

Riitahuhta, A and Aho, K (1987). Systematic engineering design as tool for improved competitiveness in the boiler industry. In: Eder, WE (Ed), ICED-87. ASME, New York, pp. 265–277.

Ringland, GA (1988). Structured object representation – schemata and frames. In: Ringland, GA and Duce, DA (Eds), Approaches to Knowledge Representation: An Introduction. Research Studies Press, pp. 81–100.

Ringland, GA and Duce, DA (1988). Approaches to Knowledge Representation: An Introduction. Wiley, New York.

Rockey, KC, Bannister, JL and Evans, HR (Eds.) (1971). Developments in Bridge Design and Construction. Crosby Lockwood, London.

Rogers, J, Feyock, S and Sobieszczanski-Sobieski, J (1988). STRUTEX – a prototype knowledge-based system for initially configuring a structure to support point loads in two dimensions. In: Gero, JS (Ed), AI in Engineering: Design, Elsevier, Amsterdam, 315–335.

Roth, K. (1987). Design models and design catalogs. In: Eder, WE (Ed), ICED-87. ASME, New York, pp. 61–67.

Rouse, WB (1978). Human problem solving in a fault diagnosis task. IEEE Transactions on Systems, Man and Cybernetics, Vol. SMC-8, 4, pp. 258–271.

Rowe, N (1988). Artificial Intelligence in PROLOG, Prentice-Hall.

Rushton, KR and Redshaw, SC (1979). Seepage and Groundwater Flow. Wiley, Chichester, 340pp.

Sacks, R and Buyukozturk, O (1987). Expert interactive design of R/C columns under biaxial bending. J Comp. Civ. Engng, 1 (2), 69–81.

Samuel, A (1963). Some studies in machine learning using the game of checkers. In: Feigenbaum, EA and Feldman, J (Eds), Computers and Thought. McGraw-Hill, New York.

Sandford, AJ (1985). Cognition and Cognitive Psychology. Weidenfeld and Nicholson, London, 435pp.

Schaffer, JD and Grefenstette, JJ (1985). Multi-objective learning via genetic algorithms. IJCAI 85. Proceedings of the Ninth International Conference on AI1 (1), pp. 58–60.

Schlaich, J (1992). Can conceptual design be taught? The Structural Engineer, 70 (9), 171–172.

Schmidt, G (1987). Expert systems in design abstraction and evaluation. In: Kalay, YE (Ed), Computability of Design. Wiley, New York, pp. 213–244.

Schon, DA (1988). Designing: rules, types and worlds. Design Studies, 9 (3), 181–190.

Serrano, D and Gossard, D (1988). Constraint management in MCAE. In: Gero, JS, AI in Engineering: Design. Elsevier, Amsterdam, pp. 217–239.

Shadbolt, N (1989). Knowledge representation in man and machine. In: Forsyth, R (Ed), Expert Systems: Principles and Case Studies, 2nd edn. Chapman and Hall, London, pp. 142–168.

Shafer, G (1976). A Mathematical Theory of Evidence. Princeton University Press, New Jersey.

Shafer, G (1982). Belief functions and parametric models. J. Roy. Stat. Soc., Series B, 44, pp. 322–352.

Shafer, G (1987). Probability judgement in artificial intelligence and expert systems. Statistical Science, 2 (1), pp. 3–16.

Shafer, G and Tversky, A (1986). Weighing evidence: the design and comparison of probability under experiments. Referenced in: Cohen, PR (1985). Heuristic Reasoning about Uncertainty: An Artificial Intelligence Approach. Pitman, London.

Shapiro, SC and Geller, J (1987). Artificial Intelligence and Automated Design. In: Kalay, YE (Ed), Computability of Design. Wiley, New York, pp. 173–187.

Shaw, MLG (1988). Problems of validation in a knowledge acquisition system using multiple experts. In: European Knowledge Acquisition Workshop, Bonn, pp. 5-2 to 5-15.

Shaw, MLG and Gaines, BR (1988). A methodology for recognising consensus. Correspondence, conflict and contrast in a knowledge acquisition system. Third AAAI Sponsored Knowledge Acquisition for Knowledge Based Systems Workshop, Banff, Canada. November, pp. 30-1 to 30-19.

Shigley, JE (1972). Mechanical Engineering Design, 2nd edn. McGraw-Hill Kogakusha, Tokyo, 753pp.

Shortliffe, EH (1976). Computer Based Medical Consultations: MYCIN, Elsevier, USA.

Shortliffe, EH and Davis, D (1979). SGART Newsletter, 55, December.

Simmons, MK (1984). Artificial intelligence for engineering design. Computer-Aided Engineering Journal, 75–83.

Simonis, H and Dincbas, M (1987). Using an extended PROLOG for digital circuit design. In: IEEE International Workshop on AI Applications to CAD Systems for Electronics, Munich, Germany, October, pp. 165–188.

Simonis, H and Dincbas, M (1990). Solving Propositional Calculus Problems in CHIP Technical Report TR-LP-46, ECRC (European Computer-Industry Research Centre), May.

Slater, P (1981). Construct systems in conflict. In: Shaw, MLG (Ed), Recent Advances in Personal Construct Technology. Academic Press, London.

Smethurst, G (1979). Basic Water Treatment. Thomas Telford, London, 216pp.

Smith, RG and Baker, JD (1983). The dipmeter advisory system: a case study in commercial expert system development. Proceedings of the Eighth International Conference on Artificial Intelligence, pp. 122–129.

Soh, AK (1985). Design of complex skeletal structures using small computers. Proceedings of the First International Conference on Education, Practice and Promotion of Computational Methods in Engineering Using Small Computers, Macao.

Soh, CK (1986). Preliminary Development of a Knowledge-Based System for the Preliminary Design of Fixed Steel Offshore Jacket Structures. MSc Thesis, Department of Civil Engineering, MIT, USA, 170pp.

Soh, CK (1990). An Approach to Automate the Design of Offshore Structures. PhD Thesis, School of Engineering, University of Wales, Cardiff, UK, 330pp.

Soh, CK and Connor, JJ (1987). Knowledge-based system for the conceptual design of fixed offshore structures. Proceedings of the Sixth ASME International Symposium on Offshore Mechanics and Artic Engineering, Houston, Texas, USAASME, New York.

Soh, CK and Miles, JC (1989). The design of steel offshore structures using an expert system. In: Topping, BHV (Ed), AI Techniques and Applications for Civil and Structural Engineers. CIVIL-COMP Press, Edinburgh, 197–201.

Soucek, B and Soucek, M (1988). Adaptive rule based and expert systems. In: Soucek, B and Soucek, M (Eds), Neural and Massively Parallel Computers – The Sixth Generation. Wiley, New York, pp. 277–289.

Speltzer, GS and Stael Von Holstein, CS (1983). Probability encoding in decision analysis. Management Science, 22 (3), pp. 340–358.

Spencer, WJ, Atkins, RM and Podlaha, P (1987). The development of an expert system for the preliminary design of bridges. Civil Engineering Systems, Vol. 6, n1/2, pp. 51–57.

Sponsler, JL (1989). Genetic algorithms applied to the scheduling of the Hubble space telescope. Telematics and Informatics, 6 (3/4), pp. 181–190.

Sriram, D (1986). Knowledge Based Approaches for Structural Design. Thesis presented to Carnegie Mellon University in Pittsburgh, Pennsylvania, February.

Sriram, D and Adey, RA (Eds.) (1987a). AI Techniques in Engineering: Tools and Techniques. Computational Mechanics Publications, Southampton, 370pp.

Sriram, D and Adey, RA (Eds) (1987b), Knowledge Based Expert Systems in Engineering: Planning and Design. Computational Mechanics Publications, Southampton, 459pp.

Stauffer, LA, Ullman, DG and Diettrich, TG (1987). Protocol analysis of mechanical engineering design. In: Eder, WE (Ed), ICED-87. ASME, New York, pp. 74–85.

Stefik, M and Bobrow, DG (1986). Object oriented programming: theories and variations. AI Magazine, January, pp. 40–62.

Stroustrup, B (1991). The C++ Programming Langauge, 2nd edn. Addison Wesley, Reading, Mass. 669pp.

Sussman, GJ (1977). Electrical design – a problem for AI research. International Joint Conference on AI, Cambridge, Massachusetts, pp. 894–900.

Sycara, K (1988). Resolving Goal Conflicts via Negotiation. AAAI-88, St Pauls, Minnesota, pp. 245–250.

Sycara, KP and Navinchandra, D (1989). Representing and Indexing Design Cases. Association for Computing Machinery, New York, pp. 735–741.

Talukdar, S, Rehy, J and Elfes, A (1988). Descriptive Models of Design Projects. In: Gero, JS (Ed), AI in Engineering: Design, Elsevier, Amsterdam, pp. 374–392.

Toll, DG, Vaptismas, N and Moula, M (1991). Comparing soils using knowledge based systems. In: Topping, BHV (Ed), AI and Civil Engineering. CIVIL-COMP Press, Edinburgh, 113–118.

Tomiyama, T, Kiriyama, T, Takeda, H, Xue, D and Yoshikawa, H (1989). Metamodel: a key to intelligent CAD systems. Research in Engineering Design, 1, 19–34.

Tong, C (1987). Towards an engineering science of knowledge based design. AI in Engineering, 2 (3), 133–166.

Trimble, G and Cooper, C (1987). Experience of knowledge acquisition for expert systems in construction. In: Proceedings of a SERC Workshop on Knowledge Acquisition for Engineering Applications. Rutherford Appleton Laboratory, Oxford, pp. 7–13.

Turing, AM (1950). Computing machinery and intelligence. Mind, 59, pp. 433–460.

Tversky, A and Kahneman, D (1977). Judgement under uncertainty: heuristics and biases. In: Johnson-Laird, PN and Wason, PC (Eds), Thinking: Readings in Cognitive Science. Cambridge University Press, Cambridge, pp. 326–340.

Ullman, DG, Stauffer, LA and Diettrich, TG (1987). Preliminary results of an experimental study of the mechanical design process. In Waldron, MB (Ed), NSF Workshop on the Design Process, Ohio State University, 143–188.

Ullman, DG, Diettrich, TG and Stauffer, LA (1988). A model of the mechanical design process based on empirical data: a summary. In: Gero, JS (Ed), AI in Engineering: Design. Elsevier, Amsterdam, pp. 193–215.

Ulrich, K and Seering, W (1987). A computational approach to conceptual design. In: Eder, W (Ed), ICED-87. ASME, New York, pp. 689–696.

Van Koppen, J (1988). A survey of expert system development tools. In: Pham, DT (Ed), Expert Systems in Engineering. ISF, Springer-Verlag, London, pp. 43–57.

Waldron, MB, Jelinek, W, Owen, D and Waldron, KJ (1987). a study of visual recall differences between expert and naive mechanical designers. In: Eder, WE (Ed), ICED-87. ASME, New York, pp. 86–93.

Wallace, KM and Hales, C (1987). Detailed analysis of an engineering design project. International Conference on Engineering Design. ICED, Boston, Massachusetts, pp. 94–101.

Wang, J and Howard, HC (1988). Design-dependent knowledge for structural engineering design. In: Gero, JS (Ed), AI in Engineering: Design. Elsevier, Amsterdam, pp. 267–278.

Waterman, DA (1986). A Guide to Expert Systems. Addison Wesley, New York.

Waterman, DA and Jenkins, B (1979). Heuristic modelling using rule based computer systems. In: Kupperman, R and Trent, D (Eds), Terrorism: Threat, Reality, Response. Hoover Institution Press, Stanford University, California, pp. 285–330.

Watson, AS (1992). Personal communication. Lecturer in Civil Engineering, Leeds University, UK.

Weizenbaum, J (1965). ELIZA – a computer program for the study of natural language communication between man and machine. Communications of ACM, 9 (1), pp. 36–45.

Weizenbaum, J (1976). Computer Power and Human Reason. Freeman, San Francisco, California.

Welbank, M (1983) A Review of Knowledge Acquisition Techniques for Expert Systems. Martlesham Consultancy Services, British Telecom Research Laboratories, Ipswich

Welbank, M (1987). Perspectives on knowledge acquisition. In: Proceedings of a SERC Workshop on Knowledge Acquisition for Engineering Applications, 1987. Rutherford Appleton Laboratory, Oxford, pp. 14–20.

Welch, J and Biswas, M (1986). Applications of expert systems in the design of bridges. Trans. Res. Rec., 1072, pp. 65–70.

Wilensky, R (1984). LISPcraft, Norton, New York.

Williams, C (1985). ART: The Advanced Reasoning Tool Conceptual Overview. Inference Corp.

Williams, T and Bainbridge, B (1988). Rule based systems. In: Ringland, GA and Duce, DA (Eds), Approaches to Knowledge Representation: An Introduction. Research Studies Press, Letchworth, pp. 101–116.

Winograd, T (1972). Understanding Natural Language. Academic Press, New York.

Winstanley, G (Ed) (1991). Artificial Intelligence in Engineering. Wiley, New York.

Winston, PH and Horn, BKP (1980). LISP, Addison Wesley, Reading, Massachusetts.

Wolchko, MJ (1987). Design by zoning code: the New Jersey office building. In: Kalay, YE (Ed), Computability of Design. Wiley, New York, pp. 273–292.

Woods, WA (1975). What's in a link?: foundations for semantic networks. In: Bobrow, D and Collins, AM (Eds), Representation and Understanding: Studies in Cognitive Science. Academic Press, New York, pp. 35–82.

Woods, WA (1983). What's important about knowledge representation? Computer, 16 (10), pp. 22–27.

Yazdani, M (1989). Building an expert system. In: Forsyth, R (Ed), Expert Systems: Principles and Case Studies, 2nd edn. Chapman and Hall, London, pp. 173–183.

Yessios, CI (1987). The computability of void architectural modelling. In: Kalay, YE (Ed), Computability of Design. Wiley, New York, pp. 141–172.

Yu, VL et al. (1979a). Evaluating the performance of a computer based consultant. In: Computer Programs in Biomedicine, 9 (1), pp. 95–102.

Yu, VL et al. (1979b). Antimicrobial selection by computer. J. Am. Med. Assoc., 242 (12), pp. 1279–1282.

Zadeh, LA (1965). Fuzzy sets. Information and Control, 8, pp. 338–353.

Zadeh, LA (1974). Fuzzy Logic and its Application to Approximate Reasoning. Information Processing. North Holland, Amsterdam, pp. 592–594.

Zadeh, LA (1975). Fuzzy logic and approximate reasoning. Synthese, 30, pp. 407–428.

Zadeh, LA (1978). Fuzzy sets as a basis of a theory of possibility. In: Fuzzy Sets and Systems. Vol. 1, pp. 3–28.

Zadeh, LA (1983). The role of fuzzy logic in the management of uncertainty in expert systems. In: Fuzzy Sets and Systems. Vol. 11, pp. 199–227.

Zhao, F and Maher, ML (1988). Using analogical reasoning to design buildings. Engineering with Computers, 4, 107–119.

Zumsteg, JR and Flaggs, DL (1985). Knowledge based analysis and design systems for aerospace structures. In Dym, CL (Ed), Applications of KBS to Engineering Design and Analysis. ASME, New York, AD-10, 67–79.

Index

Note: **Bold** type indicates a dedicated section and *italic* type indicates figures